高等职业学校"双高计划"新形态一体化教材

大数据处理技术开发应用

● 主　编　熊泽明　王兴奎

● 副主编　秦阳鸿　熊　江　余　淼

● 参　编　廖　铃　骆　伟　熊　娅　杨　勇　纪昌宁

● 主　审　陈　章　赵福奎

華中科技大學出版社

http://www.hustp.com

中国·武汉

内 容 简 介

本书主要围绕大数据处理技术展开编写，配套教学资源完善（包含录制的操作性较强的视频微课和教师授课PPT资料）。全书共分8章，第1章概述了大数据；第2章介绍了大数据平台部署的详细过程；第3章介绍了Hadoop应用开发、使用Java操作HDFS和认识MapReduce；第4章介绍了Hive数据仓库开发、Hive开发环境的搭建和Hive高级操作；第5章介绍了Flume开发应用、安装Flume、Flume自定义实现；第6章介绍了Kafka开发应用、Kafka的安装与配置、Kafka监控和编程实现；第7章介绍了PySpark开发应用、PySpark配置和PySpark案例；第8章介绍了Flink开发应用、Flink部署和Flink案例。

本书项目案例通俗易懂，大数据开发技术采用的是目前行业主流技术。实训环境部署简单，学习案例步骤完备。可作为高等职业院校大数据专业核心课程的教学用书，也可作为大数据处理技术爱好者的参考用书。

图书在版编目（CIP）数据

大数据处理技术开发应用/熊泽明，王兴奎主编. —武汉:华中科技大学出版社,2022.8

ISBN 978-7-5680-8377-5

Ⅰ.①大… Ⅱ.①熊… ②王… Ⅲ.①数据处理软件－高等职业教育－教材 Ⅳ.①TP274

中国版本图书馆CIP数据核字(2022)第127179号

大数据处理技术开发应用
Dashuju Chuli Jishu Kaifa Yingyong

熊泽明　王兴奎　主编

策划编辑：张　玲　徐晓琦
责任编辑：陈元玉
封面设计：原色设计
责任监印：周治超
出版发行：华中科技大学出版社(中国·武汉)　电话：(027)81321913
武汉市东湖新技术开发区华工科技园　邮编：430223
录　排：武汉金睿泰广告有限公司
印　刷：武汉市籍缘印刷厂
开　本：787mm × 1092mm　1/16
印　张：15.5
字　数：362千字
版　次：2022年8月第1版第1次印刷
定　价：55.00元

Preface 前 言

21世纪，随着现代信息技术的不断发展，世界已跨入了互联网+大数据时代。大数据产业正在深刻改变着人们的思维、生产和生活方式，正在掀起新一轮的产业和技术革命。大数据技术历经“十三五”期间的孕育成长后，目前大数据已覆盖政府、金融、交通、企业、教育、医疗等各应用领域，与5G通信技术、物联网技术、互联网产业相融合，在大数据技术领域起着重要的支撑作用。特别是在2020年疫情以后，大数据技术这个词已是家喻户晓，其应用极为火爆，为人们的衣、食、住、行提供服务。目前各家企业都有自己发行的大数据版本。虽然各家企业的大数据处理技术都由自己的研发团队设计，但目前主流的大数据都是基于开源技术的Hadoop大数据平台进行开发与运维的。在从事大数据技术运维与管理的工作中，都是围绕着开源Hadoop系统核心技术去开展工作的。在大数据领域，很多核心技术都是基于开源Hadoop系统的。

本书在编写过程中，主要围绕大数据处理技术生态圈展开。将大数据平台运维教学与企业大数据开发实战运维工作相结合，将目前主流的大数据运维技术整合为大数据综合实训案例知识点，适当融入课程思政的内容，对本书难点、重点部分录制了操作性较强的视频微课，形成一本实操性较强的大数据处理技术专业书籍。读者能够快速了解大数据处理技术和大数据底层开发核心技术，通过理论+综合实训方法，快速掌握目前大数据的核心知识点和技能点。通过本书和社区技术的结合，能够快速提升读者的自学能力，熟练掌握目前主流的大数据处理技术。

本书为大数据专业核心课程用书，所涉及的大数据处理技术仅限于教学和读者学习使用，不用于任何商业活动。本书由重庆三峡职业学院的熊泽明教授、北京华晟经世信息技术股份有限公司的王兴奎工程师担任主编，由重庆三峡职业学院的熊江教授及余淼副教授、秦

阳鸿担任副主编。重庆三峡职业学院的骆伟副教授、廖铃、熊娅、杨勇及纪昌宁高级实验师等参与部分内容的编写及审校工作。在编写过程中，我们得到了业内部分大数据相关企业及工程师的支持和帮助，引用了互联网中的大量资料（包括文本和图片等），核心技术来自大数据技术社区官方帮助文档，在此深表谢意。由于编者能力有限，书中难免存在不足之处，望广大读者不吝赐教。

编 者

2022年3月

Contents 目录

第1章 大数据概述

第2章 Hadoop平台部署

第 4 章 Hive数据仓库开发

第 5 章 Flume开发应用

第 6 章 Kafka开发应用

第7章 PySpark开发应用

第8章 Flink开发应用

第1章　大数据概述

学习目标

（1）了解大数据的起源和含义、大数据常用开发工具Hadoop。

（2）理解大数据的特征和大数据思维的原理。

（3）理解大数据的行业发展趋势。

（4）掌握大数据应用的开发流程和大数据采集方法。

思政目标

（1）了解国家“十四五”规划对大数据产业的要求。

（2）了解大数据对民生发展的重要性。

1.1　大数据简介

1.1.1　大数据的发展历程

说“我们生活在一个网络时代”，显得有点落伍了，当下最时髦的说法是“我们生活在一个大数据时代”。在我国，大数据已经被写入“十四五”规划中。那如何来理解大数据呢？从字面上看，人们是用“大数据”来描述和定义信息爆炸时代产生的海量数据。但实际上，“大数据”的渗透能力远远超出人们的想象，不管是在物理学、生物学、环境生态学等领域，还是在军事、金融、通信、贸易等行业，数据正在迅速膨胀，几乎所有领域都被波及。“大数据”正在改变，甚至颠覆我们所处的时代，对社会发展产生的影响，也让我们的思维不得不跟随时代的变迁而经历自我革命。

大数据本身的发展也可以分为三个阶段。

第一个阶段是萌芽期。这一阶段随着数据挖掘理论和数据库技术的逐步成熟，一批商业智能工具和知识管理技术开始被应用，如数据仓库、专家系统、知识管理系统等，企业、机构对内部数据进行统计、分析和挖掘利用。

第二个阶段是成熟期。这一阶段非结构化数据大量产生，传统处理方法难以应对。因此，带动了大数据技术的快速发展，大数据解决方案逐渐走向成熟，形成并行计算与分布式系统两大核心技术，谷歌公司的GFS和MapReduce等大数据技术受到追捧，Hadoop平台开始大行其道。

第三阶段是大规模应用期。大数据应用渗透各行各业，数据驱动决策，信息社会智能化程度大幅提高，同时将出现跨行业、跨领域的数据整合，甚至是全社会的数据整合，从各种各样的数据中找到对社会治理、产业发展更有价值的信息。

工业和信息化部于2021年11月30日印发《“十四五”大数据产业发展规划》。到2025年，大数据产业测算规模突破3万亿元，年均复合增长率保持在25%左右，创新力强、附加值高、自主可控的现代化大数据产业体系基本形成。该规划提出，推动建立市场定价、政府监管的数据要素市场机制，发展数据资产评估、登记结算、交易撮合、争议仲裁等市场运营体系。培育大数据交易市场，鼓励各类所有制企业参与要素交易平台的建设，探索多种形式的数据交易模式。这些标志着我国将大力发展大数据应用在多领域内的发展。

1.1.2 大数据的特征

大数据（big data）研究机构Gartner给出了这样的定义。大数据是一种需要新处理模式才具有更强的决策力、洞察发现力和流程优化能力,以适应海量、高增长率和多样化的信息资产。麦肯锡全球研究所给出的定义是：一种规模在获取、存储、管理、分析等方面远远超出传统数据库软件工具能力范围的数据集合，具有海量的数据规模、快速的数据流转、多样的数据类型和价值密度低等四大特征。

我国工业和信息化部于2021年11月30日印发的《“十四五”大数据产业发展规划》中将大数据定义为：大数据是数据的集合，以容量大、类型多、速度快、精度准、价值高为主要特性（见图1-1），是推动经济转型发展的新动力，是提升政府治理能力的新途径，是重塑国家竞争优势的新机遇。

1.容量

顾名思义，大数据的5个特性中的主要特性为容量大。大数据中数据的采集、存储和计算的量都非常大。正常的计算机处理4G数据需要约4分钟，处理1 TB的数据需要3个小时，而处理1 PB的数据需要4个月零3天。起始计量单位只有达到PB的数据才可以称为大数据。

淘宝网是当今较早开始投资和部署大数据应用的中国企业之一，也因此从中获利。大数据让淘宝网改变了重复销售的策略，这带来了10%~15%在线销售的明显涨幅，增加收入10亿美元。淘宝网自己有一个庞大的大数据生态系统。淘宝网每小时完成约100万笔交易，大数据生态系统每天会处理TB级的新数据和PB级的历史数据，以及分析数以百万计的产品数据、数以亿计的客户和搜索关键词。

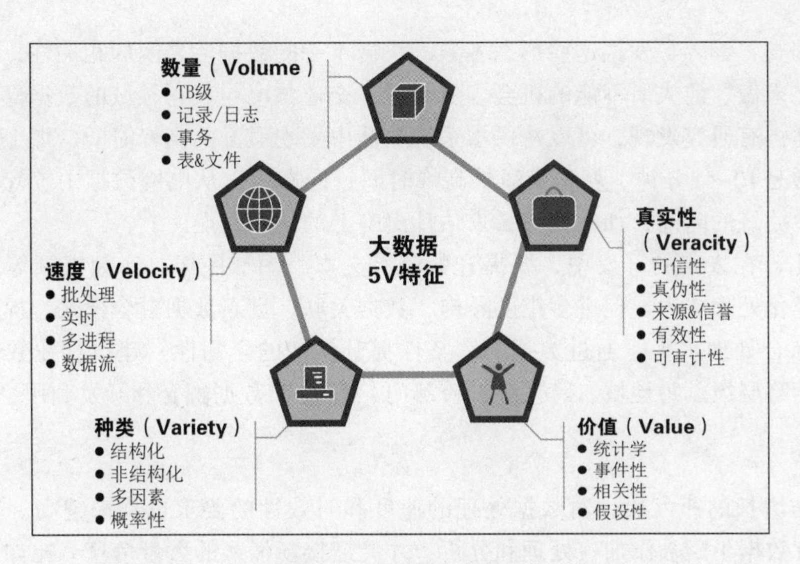

图1-1 大数据的5V特性

2. 类型

大数据不仅体现在量的急剧增长上，数据类型亦是多样，可分为结构化数据、半结构化数据和非结构化数据。多年来，结构化数据一直存储在关系型数据库中；半结构化数据包括电子邮件、文字处理文件以及大量的网络新闻等，以内容为基础，这也是谷歌和百度存在的理由；而非结构化数据随着社交网络、移动计算和传感器等新技术的不断产生，广泛存在于社交网络、物联网和电子商务中。

深究下去，非结构化事实上是未必成立的概念。信息里的“结构”是永远存在的，只不过结构尚未被发现，或者结构变化不定（半结构化或多结构化），或者结构存在但机器处理不了，就像典型的非结构化数据，即文本一样，它有语言学意义上的结构（语法和语义），又有叙事意义上的结构（三段式、先破后立等），还有结构化的元数据（作者、标题、发布时间等），但文本一直是非结构化数据的典型。有学者说：非结构化？此言差矣；应该说非模型化（unmodeled），结构本在，只是未建模而已。早期的非结构化数据，在企业的数据语境里主要是文本，如电子邮件、文档、健康/医疗记录。随着互联网和物联网的发展，又扩展到网页、社交媒体、感知数据，涵盖音频、图片、视频、模拟信号等，真正诠释了数据的多样性。

从另一个维度上看，数据的多样性又表现在数据的来源和用途上。就卫生保健数据来讲，大致有药理学科研数据、临床数据、个人行为和情感数据、就诊/索赔记录和开销数据四类。麦肯锡在《大数据：创新、竞争和生产力的下一个前沿》里关于美国卫生保健行业如何利用多样化数据给出了很好的建议，有兴趣的读者可以去阅读此书。

又如交通领域。北京市交通智能化分析平台数据源来自路网摄像头/传感器、地面公交、轨道交通、出租车以及省际客运、旅游、化危运输、停车、租车等运输行业，还有问卷调查和GIS数据。从数据体量和速度上也达到了大数据的规模：4万辆浮动车每天产生2000万条记录；交通卡刷卡记录每天1900万条；手机定位数据每天1800万条；出租车运营数据每天100万条；高速ETC数据每天50万条；针对8万户家庭的定期调查，等等。发掘这些形态各异、快慢不一的数据流之间的相关性，是大数据做前人之未做、前人所不能的机会。更甚者，交通状况与其它领域的数据都存在较强的关联性：有研究发现，可以从供水系统数据中发现晨洗的高峰时间，加上一个偏移量（通常是40~45分钟）就是交通早高峰时间；同样可以从电网数据中统计出傍晚办公楼集中关灯的时间，加上偏移量来估计出晚上的堵车时点。

因此，在未来的开发中，要提升数值、文本、图形图像、音频视频等多类型数据的多样化处理能力。促进多维度异构 数据关联，创新数据融合模式，提升多模态数据的综合处理水平，通过数据的完整性提升认知的全面性。建设行业数据资源目录，推动跨层级、跨地域、跨系统、跨部门、跨业务 数据融合和开发利用。

3. 速度

数据增长的速度快，对数据处理的速度和时效性的要求也越来越高。数据来源广，只有数据不断被添加、处理和分析，才能迎接新涌入的大量信息，比如，在搜索引擎中，用户要能够查询到几分钟前的新闻，个性化推荐算法尽可能满足实时推荐的要求。这是大数据区别于传统数据的显著特征。网络时代，通过高速的计算机和服务器，创建实时的数据流已成为主流趋势。企业不仅需要了解如何快速创建数据，还需要如何快速处理、分析数据并返回给用户，以满足用户的实时需求。

4. 精度

数据的精度高，即数据的质量高。数据本身如果是虚假的，那么它就失去了存在的意义，因为任何通过虚假数据得出的结论可能是错误的，甚至是相反的。

5. 价值

相比于传统的小数据，大数据的最大价值在于，通过从各种类型的数据中挖掘出对未来趋势与模式预测分析有价值的数据，再通过机器学习方法、人工智能方法或数据挖掘方法进行深度分析，发现新规律和新知识,并运用于农业、金融、医疗等各个领域，从而达到改善社会、提高生产率、推进科学研究的效果。但是，大数据的价值密度低，约80%的数据都是无效数据，比如，在视频监控过程中，可能获取到的有用数据只有一两秒。因此，如何使用数据分析与挖掘算法快速提取出数据中的有效信息，是当今面临的研究难题。

数据的价值，归根到底只有使用，才能发挥其作用。从数据的使用方式来看，数据一般可以分为以下两种。对内：为企业发展决策做支撑，帮助企业更高效地制定策略；支持一线营销管理工作，对目标客户进行精准营销，拓展业务。对外：实现数据的长尾效应，对数据进行整合、抽取，与客户进行合作，发挥数据的外在作用。

以传统运营商为例，数据的价值为：大数据将会成为推动人类社会发展的“新石油”，也将成为未来提高企业竞争力的关键要素。正确认识和发挥大数据的作用，不夸大、不缩小，是有效应用大数据的前提。可以预见的是，大数据就在你我身边，不论你是否察觉，它都会变成你生活的一部分。大数据时代，挑战与机遇并存。正确处理与利用好大数据，是企业得以发展的必备动力。只有更好地发挥数据的作用，才能在大数据时代更好、更快地提升竞争力，进而引领整个行业的发展。

1.1.3 大数据思维

大数据时代开起了数据思维的创新和变革:多、杂、好。数据思维就是根据数据去思考事物的一种思维模式，这种模式往往对真实比较重视，也就是追求真理。现在企业越来越依靠数据，根据数据分析找出问题，然后跟踪问题进行问题的解决。数据化思维的内涵不是新鲜事物，只是说我们对思维的认知有了一个新的解读和认知。学会利用所有的数据，不能依赖于部分数据。接受不准确性。其实在混乱中更能代开自己的思维，发现新的世界。不必要知道现象背后的具体原因，学会用数据说话。要有一定的逻辑思维和洞察力。

《“十四五”大数据产业发展规划》中强调，加强大数据知识普及，通过媒体宣传、论坛展会、赛事活动、体验中心等多种方式，宣传产业典型成果，提升全民大数据认知水平。加大对大数据理论知识的培训，提升全社会获取数据、分析数据、运用数据的能力，增强利用数据创新各项工作的本领。推广首席数据官制度，强化数据驱动的战略导向，建立基于大数据决策的新机制，运用数据加快组织变革和管理变革。

1.2 大数据应用开发流程

大数据中，不仅数据处理规模巨大，而且数据处理需求多样化，超出了当前计算机存储和处理的能力，因此，数据处理能力成为核心竞争力。大数据处理流程主要包括数据采集、数据预处理、数据存储、数据分析与挖掘、数据展示与应用等环节。

1.2.1 数据采集

世界上每时每刻都在产生大量的数据，包括物联网传感器数据、社交网络数据、商品交易数据等。面对如此大的数据，与之相关的采集、存储、分析等环节产生了一系列问题。如何收集这些数据，并且进行转换、分析、存储及有效已成为很大挑战。需要有这样一个系统来收集这样的数据，并且对数据进提取、转换和加载。

本节就介绍这样一种大数据采集技术。什么是大数据采集技术？大数据采集技术就是对数据进行ETL操作，通过对数据进行提取、转换和加载，最终挖掘数据的潜在价值，然后提供给用户解决方案或者决策参考。ETL，是英文Extract-Transform-Load的缩写，数据从数据来源端经过提取（extract）、转换（transform）、加载（load）到目的端，然后进行分析的过程。用户从数据源抽取出所需的数据，经过数据清洗,按照预先定义好的数据模型将数据加载到数据仓库中去，最后对数据仓库中的数据进行数据分析和处理。数据采集位于数据分析生命周期的重要一环，它通过传感器数据、社交网络数据、移动互联网数据等方式获得各种类型的结构化、半结构化及非结构化的海量数据。由于采集的数据种类错综复杂，对于这种不同种类的数据分析，必须通过提取技术。将复杂格式的数据进行数据提取，从数据原始格式中提取出我们需要的数据，这里可以丢弃一些不重要的字段。对于提取后的数据，由于数据源头的采集可能存在不准确性，所以我们必须进行数据清洗，对于那些不正确的数据进行过滤、剔除。对数据进行分析的工具或者系统不同，我们还要对数据进行转换操作，将数据转换成不同的数据格式，最终按照预先定义好的数据仓库模型将数据加载到数据仓库中去。

实际开发中，数据产生的种类很多，并且不同种类的数据产生的方式不同。大数据采集系统主要分为以下三类。

1. 系统日志采集系统

许多公司的业务平台每天都会产生大量的日志信息。关于这些日志信息，我们可以得出很多有价值的数据。通过对这些日志信息进行日志采集、收集，然后进行数据分析，挖掘出公司业务平台日志数据中的潜在价值，可为公司决策和后台服务器性能评估提供可靠的数据保证。系统日志采集系统的任务就是收集日志数据、提供离线和在线的实时分析。目前常用的开源日志收集系统有Flume、Scribe等。Apache Flume是一个分布式的、可靠性的、可用的数据收集系统，用于高效地收集、聚合和移动大量的日志数据，具有基于流式数据流的简单灵活的架构。其可靠性机制、故障转移和恢复机制，使得Flume具有强大的容错能力。Scribe是Facebook开源的日志采集系统。Scribe实际上是一个分布式共享队列，它可以从各种数据源上收集日志数据，然后放入其上面的共享队列中。Scribe可以接受Thrift客户端发送过来的数据，将其放入其上面的消息队列中。然后通过消息队列将数据Push到分布式存储系统中，并且由分布式存储系统提供可靠的容错性能。如果最后的分布式存储系统崩溃，Scribe中的消息队列还可以提供容错能力，还会将日志数据写入本地磁盘中。Scribe支持持久化的消息队列来提供日志收集系统的容错能力。

2. 网络数据采集系统

通过网络爬虫和一些网站平台提供的公共API(如Twitter和新浪微博API)等从网站上获取数据。这样就可以将非结构化和半结构化的网页数据从网页中提取出来，并将其进行清洗、转换成结构化的数据，将其存储为统一的本地文件数据。目前常用的网络爬虫系统有Apache Nutch、Crawler4j、Scrapy等框架。Apache Nutch是一个高度可扩展的和可伸缩性的分布式爬虫框架。Apache通过分布式抓取网页数据，由Hadoop系统提供支持，通过提交MapReduce任务来抓取网页数据，并将网页数据存储在分布式文件系统中。Nutch主要用于收集网页数据，然后对其进行分析、建立索引，再提供相应的接口来对其网页数据进行查询的一套工具。由于多台机器并行执行爬取任务，Nutch充分利用多台机器的计算资源和存储能力，大大提高了系统爬取数据的能力。Crawler4j、Scrapy都是爬虫框架，给开发人员提供便利的爬虫API接口。开发人员只需要关心爬虫API接口的实现，不需要关心框架是怎么爬取数据的。Crawler4j、Scrapy框架大大降低了开发人员的开发速率，开发人员可以很快完成一个爬虫系统的开发。

3.数据库采集系统

一些企业会使用传统的关系型数据库如MySQL和Oracle等来存储数据。除此之外，Redis和MongoDB这样的NoSQL数据库也常用于数据的采集工作。企业每时每刻产生的业务数据，将以数据库一行记录的形式被直接写入数据库中。通过数据库采集系统直接与企业业务后台服务器结合，将企业业务后台每时每刻产生的大量业务记录写入数据库中，最后由特定的处理分析系统进行系统分析。

目前主要流行以下大数据采集分析技术。Hive是Facebook团队开发的一个可以支持PB级别的可伸缩性的数据仓库。这是一种建立在Hadoop之上的开源数据仓库解决方案。Hive支持使用类似SQL的声明性语言（如HiveQL）表示的查询，这些语言被编译为使用Hadoop执行的MapReduce作业。另外，HiveQL语言可使用户将自定义的map-reduce脚本插入查询中。该语言支持基本数据类型，类似数组和Map的集合以及嵌套组合。HiveQL语句被提交执行。首先，Driver将查询语句传递给编译器，通过典型的解析、类型检查和语义分析，使用存储在MetaStore中的元数据。编译器生成一个逻辑任务，然后通过简单的基于规则的优化器进行优化，最后生成一组MapReduce任务和由HDFSTask的DAG优化后的任务。然后使用Hadoop系统按照其依赖性顺序执行这些任务。Hive降低了对那些不熟悉Hadoop MapReduce接口的用户学习门槛，Hive提供了一些简单的HiveQL语句，对数据仓库中的数据进行简要分析与计算。

在大数据采集技术中，其中一个关键环节就是转换操作。它将清洗后的数据转换成不同的数据形式，由不同的数据分析系统和计算系统进行处理与分析。将批量数据从生产数据库加载到Hadoop分布式文件系统中或者从Hadoop分布式文件系统将数据转换为生产数据库中，这是一项艰巨的任务。用户必须考虑确保数据的一致性，生产系统资源的消耗等问题。使用脚本传输数据的效率低下且耗时。Apache Sqoop就是用来解决这个问题的，Sqoop允许从结构化数据存储（如关系数据库、企业数据仓库和NoSQL系统）轻松导入和导出数据。使用Sqoop，可以将来自外部系统的数据配置到HDFS上，并将表填入Hive和HBase中。运行Sqoop时，被传输的数据集被分割成不同的分区，一个只有mapperTask的Job被启动，mapperTask负责传输这个数据集的一个分区。Sqoop使用数据库元数据来推断数据类型，因此，每个数据记录都以类型安全的方式进行处理。

1.2.2 数据预处理

由于大数据所要进行分析的数据量迅速膨胀（已达G或T数量级），同时，由于各种导致现实世界数据集中常包含噪声（数据中存在的错误或异常的数据）、不完整（感兴趣的属性没有值）、甚至是不一致(数据内出现不一致的情况)的数据，如图1-2所示。

必须对数据挖掘所涉及的数据对象进行预处理。数据预处理主要包括：数据清洗、数据集成、数据转换、数据归约。预处理是数据挖掘（知识发现）过程中的一个重要步骤，尤其对包含噪声、不完整，甚至是不一致的数据挖掘时，更需要进行数据的预处理，以提高数据挖掘对象的质量，并最终达到提高数据挖掘所获模式质量的目的。

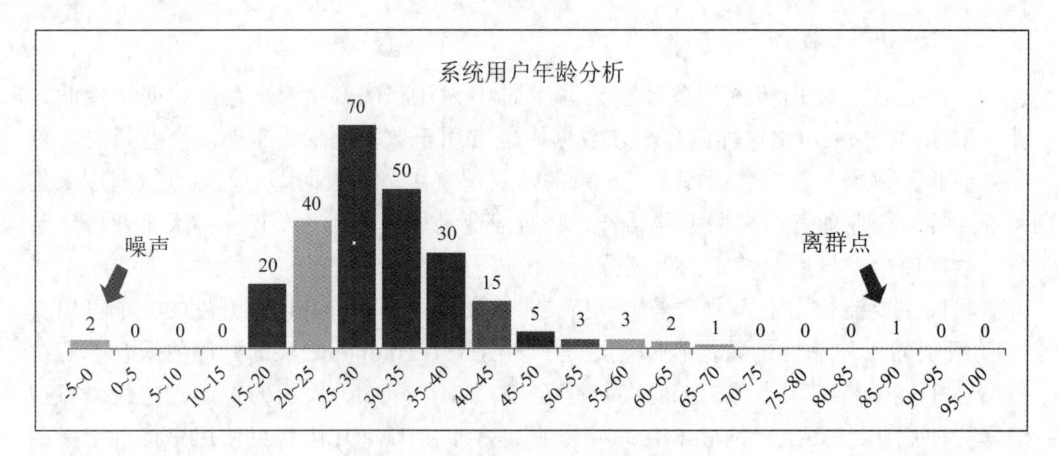

图1–2 数据统计中包含的噪声数据

1. 数据清洗

数据清洗是指消除数据中所存在的噪声；数据集成是指将来自多个数据源的数据合并到一个完整的数据集；数据转换是指将一种格式的数据转换为另外一种格式的数据；数据归约是指通过删除冗余特征来消除多余数据。包含不完整、有噪声和不一致的数据对大规模数据库来讲是非常普遍的情况。不完整数据的产生有以下几个原因。

（1）有些属性的内容有时没有。

（2）有些数据当时被认为是不必要的。

（3）由于误解或检测设备失灵导致相关数据没有记录下来。

（4）与其他记录的内容不一致而被删除。

（5）历史记录或对数据的修改被忽略了。

遗失数据，尤其是一些关键属性的遗失数据或许需要推导出来。噪声数据的产生原因有以下几个。

（1）数据采集设备有问题。

（2）数据录入过程中发生了人为或计算机错误。

（3）数据传输过程中发生错误。

（4）由于命名规则和数据代码不同而引起的不一致。数据清洗还将删除重复记录行。

2. 数据集成

数据集成就是将来自多个数据源（如数据库文件等）的数据合并到一起。由于描述同一个概念的属性在不同的数据库中取不同的名字，所以在进行数据集成时常会引起数据的不一致和冗余。命名的不一致也常会导致同一属性值的内容不同。同样，大量的数据不仅会降低挖掘速度，而且会误导挖掘进程。因此，除进行数据清洗之外，在数据集成中还需要消除数据的冗余。此外，在完成数据集成之后，有时还需要进行数据清洗，以便消除可能存在的数据冗余。

3. 数据转换

数据转换主要是对数据进行规范化操作。在正式进行数据挖掘之前，尤其是使用基于对象距离的挖掘算法时，如神经网络、最近邻分类等，必须进行数据规格化。也就是将其缩至特定的范围内。数据转换也可以是属性的生成，也就是利用当前存在的一个或多个属性生成另外一个有意义的属性。

4. 数据规约

数据规约的目的就是缩小所挖掘数据的规模，但却不会影响（或基本不影响）最终的挖掘结果。现有的数据归约方法包括以下几种。

（1）数据聚合。

（2）削减维数，例如，通过相关分析，消除多余属性。

（3）数据压缩，例如，利用聚类或参数模型替代原有数据。此外，利用基于概念树的泛化也可以实现对数据规模的削减。

这里需要强调的是，以上所提及的各种数据预处理方法，并不是相互独立的，而是相互关联的。由于所要分析的数据是含有噪声、不完全和不一致性的，数据预处理能够帮助改善数据的质量，进而帮助提高数据挖掘进程的有效性和准确性。高质量的决策来自高质量的数据。因此，数据预处理是整个数据挖掘与知识发现过程中的一个重要步骤。

1.2.3 数据存储

大数据因为规模大、类型多样、增长速度快，所以在存储和计算上，都需要技术支持，依靠传统的数据存储和处理工具，已经很难实现高效的处理了。

以往的数据存储，主要是基于关系型数据库。而关系型数据库，在面对大数据的时候，存储设备所能承受的数据量是有上限的，当数据规模达到一定的量级之后，数据检索的速度就会急剧下降，对于后续的数据处理来说，也带来了困难。为了解决这个主题，主流的数据库系统都纷纷给出解决方案，比如，MySQL提供了MySQL proxy组件，实现了对请求的拦截，结合分布式存储技术，从而可以将一张很大表中的记录拆分到不同的节点上去进行查询。对于每个节点来说，数据量不会很大，从而提升了查询效率。

1. 实时数据库查询模式

传统关系型数据库基于存储模式带来的存储和访问瓶颈问题，是无法靠自身解决的，于是有了基于Big-Table型的NoSQL数据库的用武之地。比较典型的技术组合就是HDFS+HBase，利用HDFS的分布式、高可用数据存储，结合HBase面向列的数据存储模型来从而解决大数据量存储的问题；结合HBase基于Rowkey自然序的存储来实现海量数据快速查询。当然，这种模式只适用于结构型数据，且只适用于历史数据查询，不适用于事务型业务的处理。

2. 数据仓库

基于关系型数据库的数据仓库，同样面临数据存储规模的问题。因此，在银行业务中，同样也只能存储短期的数据，其目标在于支持基于业务年度的报表统计和业务分析，而对于超过一定期限的数据，仍然使用数据磁盘或磁带存储的模式。基于大数据技术体系，采用HDFS+Hive的模式，构建大数据仓库，可以解决数据大基数存储的问题。

1.2.4 数据分析

数据分析是大数据处理与应用的关键环节。它决定了大数据集合的价值和可用性，以及分析预测结果的准确性。通过数据采集、预处理以及存储环节，从异构的数据源中获得用于大数据处理的原始数据，用户可以根据自己的需求对这些数据进行处理与存储，在数据分析与挖掘环节，根据大数据的应用情境与决策需求，选择合适的数据分析技术，将有价值的知识或重要信息从数据库的相关数据集合中提取出来，提高大数据分析结果的可用性、价值和准确性。该流程主要分为四个方向：可视化分析、数据挖掘、预测性分析、语义分析。

1. 可视化分析

由于人类大脑的记忆能力有限，所以我们利用视觉获取的信息量多于感官，在大数据与互联网时代，企业从传统的流程式管理方式过渡到基于数据的管理方式将成为必然趋势，数据可视化能够帮助分析的人对数据有更全面的认识。让图形替数据说话：不管是对数据分析专家还是对普通用户，数据可视化是数据分析工具最基本的要求。可视化可以直观地展示数据，让数据说话，让观众直接“听到”结果。常用的方法有以下几种。

（1）在数据采集过程中进行数据分类，根据数据属性和方法去解决问题。

（2）可视化映射：将数据的数值、空间坐标、不同位置数据间的联系等映射为可视化视觉通道的不同元素，如标记、位置、形状、大小和颜色。

（3）数据转换和处理：通过去噪，清洗数据，提取数据。

（4）用户验证：数据的正确与否，需要用户的大胆假设和积极验证，反复验证数据的合理性等，从而向公众或者上司展示数据。

2. 数据挖掘

可视化面向的对象是人，数据挖掘则是面向机器。集群、分割、孤立点分析及其他算法让我们深入数据内部，挖掘价值。数据挖掘是从大量的、不完全的、有噪音的、模糊不清的、随机的实际数据中提取出蕴含其中的、大家事前不清楚的、具备潜在有效性的信息和知识的过程。

开展数据挖掘的数据源必须是真实的，因为这些数据源可能不完整或者包含一些干扰数据项。发现的信息和知识必须是用户感兴趣的和有用的。一般来讲，数据挖掘的结果并不要求是完全正确的知识，只是一种趋势。

数据挖掘涉及的知识面广，技术点多。在面对复杂多样的业务分析场景时，如何制定有效的数据挖掘分析方案，实际上有一套可遵循的方法体系。数据挖掘的技术有很多种，下面着重讨论常用的数据挖掘技术：统计技术、关联规则、基于历史的分析、遗传算法、聚集检测、连接分析、决策树、神经网络、粗糙集、模糊集、回归分析、差别分析、概念描述等。

1）统计技术

数据挖掘涉及的科学领域和技术很多，如统计技术。统计技术对数据集进行挖掘的主要思想是：统计方法对给定的数据集合假设了一个分布或者概率模型（例如正态分布），然后根据模型采用相应的方法来进行挖掘。

2）关联规则

数据关联是数据库中存在的一类重要的可被发现的知识。若两个或多个变量的取值之间存在某种规律性，就称为关联。关联可分为简单关联、时序关联和因果关联。关联分析的目的是找出数据库中隐藏的关联网。有时并不知道数据库中数据的关联函数，即使知道，也是不确定的，因此关联分析生成的规则带有可信度。

3）基于历史的（Memory-Based Reasoning，MBR）分析

先根据经验知识寻找相似的情况，然后将这些信息应用于当前的例子中。这就是MBR的本质。MBR首先寻找和新记录相似的邻居，然后利用这些邻居对新数据进行分类和估值。使用MBR有三个主要问题，寻找确定的历史数据；确定历史数据的最有效方法；确定距离函数、联合函数和邻居的数量。

4）遗传算法

基于进化理论，并采用遗传结合、遗传变异以及自然选择等设计方法的优化技术。遗传算法的主要思想是：根据适者生存的原则，形成由当前群体中最适合的规则组成新的群体，以及这些规则的后代。典型情况下，规则的适合度（Fitness）用它对训练样本集的分类准确率进行评估。

5）聚集检测

将物理或抽象对象的集合分组成为由类似的对象组成的多个类的过程称为聚类。由聚类所生成的簇是一组数据对象的集合，这些对象与同一个簇中的对象相似，与其他簇中的对象相异。相异度是根据描述对象的属性值来计算的，经常采用的度量方式是距离。

6）连接分析

连接分析（Link Analysis）的基本理论是图论。图论的思想是寻找一个可以得出好结果但不是完美结果的算法，而不是去寻找完美的解的算法。连接分析就是运用了这样的思想：不完美的结果如果是可行的，那么这样的分析就是一个好的分析。利用连接分析，可以从一些用户的行为中分析出一些模式；同时将产生的概念应用于更广的用户群体中。

7）决策树

决策树提供了一种展示类似在什么条件下会得到什么值这类规则的方法。

8）神经网络

结构上，可以把神经网络划分为输入层、输出层和隐含层。输入层的每个节点对应一个个的预测变量。输出层的节点对应目标变量，可有多个。在输入层和输出层之间是隐含层（对神经网络使用者来说不可见），隐含层的层数和每层节点的个数决定了神经网络的复杂度。

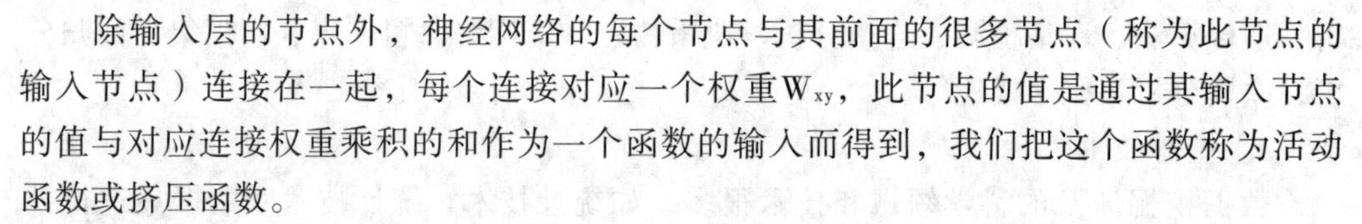

除输入层的节点外，神经网络的每个节点与其前面的很多节点（称为此节点的输入节点）连接在一起，每个连接对应一个权重W_{xy}，此节点的值是通过其输入节点的值与对应连接权重乘积的和作为一个函数的输入而得到，我们把这个函数称为活动函数或挤压函数。

9）粗糙集

粗糙集理论基于给定训练数据内部的等价类的建立。等价类内部的数据元组是不加区分的，即对于描述数据的属性，这些样本是等价的。给定现实世界数据，有些类不能被可用的属性加以区分。粗糙集就是用来近似或粗略地定义这种类。

10）模糊集

模糊集理论是将模糊逻辑引入数据挖掘分类系统，允许定义“模糊”域值或边界。模糊逻辑使用0.0和1.0之间的真值来表示一个特定的值是一个给定类成员的隶属程度，而不是用类之间的精确截断。模糊逻辑提供了在高抽象层处理的便利。

11）回归分析

回归分析分为线性回归、多元回归和非线性同归。在线性回归中，数据用直线建模，多元回归是线性回归的扩展，涉及多个预测变量。非线性回归是在基本线性模型的基础上添加多项式而形成的非线性同门模型。

12）差别分析

差别分析的目的是试图发现数据中的异常情况，如噪音数据、欺诈数据等异常数据，从而获得有用信息。

13）概念描述

概念描述就是对某类对象的内涵进行描述，并概括这类对象的有关特征。概念描述分为特征性描述和区别性描述，前者描述某类对象的共同特征，后者描述不同类对象之间的区别，生成类的特征性描述只涉及该类中所有对象的共性。

由于人们急切需要将存在于数据库和其他信息库中的数据转化为有用的知识，因此数据挖掘被认为是一个新兴的、非常重要的、具有广阔应用前景和富有挑战性的研究领域，并引起了众多学科（如数据库、人工智能、统计学、数据仓库、在线分析处理、专家系统、数据可视化、机器学习、信息检索、神经网络、模式识别、高性能计算机等）研究者的广泛关注。作为一门新兴的学科，数据挖掘是由上述学科相互交叉、相互融合而成的。

3. 预测性分析

预测性分析涵盖了各种统计学技术，主要利用预测模型、机器学习、数据挖掘等技术来分析当前的及历史的数据，从而对未来或其他不确定的事件进行预测。

在商业领域，预测模型从历史和交易数据中寻找规律，用来识别可能的风险和商机。预测模型可以捕捉各个因素之间的联系，以评估风险及与之相关的潜在条件，从而指导交易方案的决策。

这些技术的功能和效应是指，预测分析为每一个个体（比如客户、员工、患者的医疗、产品SKU、车辆、部件、机器或其他组织单位）（以概率的形式）提供预测评分，从而决策、反馈或影响针对上述个体的组织性流程。这些流程包括营销、信用风险评估、欺诈检测、制造、医疗保健、政府的运作，甚至执法。

预测性分析方法被广泛应用于精算科学、营销、金融服务、保险、电信、零售、旅行、保健、制药、能力规划及其他领域。

其中最著名的应用是信用评分，这项应用贯穿了整个金融服务体系。评分模型用于处理客户的信用记录、贷款申请、客户数据等，从而分析个体（客户）在未来的还贷可能性，并依照分析结果将客户进行排序。

4. 语义分析

在现实世界中，事物所代表的概念，以及它与其他概念之间的关系，可以被认为是语义。语义是对符号的解释，比如苹果是一种水果，含有丰富的矿物质和维生素。解释了“苹果”这个字符串（符号、概念）。

大数据4V特征中的一个典型特征是Variety，它有多方面的含义，最主要的是指数据类型的多样化。当表示一本图书时，可以有数值型、日期型、文本型等多种类型。

在“数据存储”中我们介绍了结构化数据的表示方法，以下表示了两本书B1、B2。

- B1:（互联网大数据处理技术与应用，曾剑平，清华大学出版社，2017，大数据类）。
- B2:（数学之美，吴军、人民邮电出版社，2014，数学类）。

在图书推荐的大数据应用中，要决定把什么书推荐给客户，最基本的问题是计算两本书的相似度。就B1、B2这两本书来说，出版年份2017和2014之间的相似度就比较容易计算，但是“大数据类”和“数学类”简单依靠字符串就无法准确计算了，“互联网大数据处理技术与应用”和“数学之美”等文本型的就更难定了。而这些问题在大数据分析及应用中是非常普遍的，因此，对于大数据分析应用而言，语义分析计算的重要性不言而喻，直接影响到最终的大数据价值。

1.2.5 数据可视化

数据是对客观事物的性质、状态以及相互关系等进行记载的物理符号或这些物理符号的组合。它是可识别的、抽象的符号。数据不仅指狭义上的数字，还可以是具有一定意义的文字、字母、数字符号的组合、图形、图像、视频、音频等，也是客观事物的属性、数量、位置及其相互关系的抽象表示。如“0、1、2……”、“阴、雨、下降、气温”、“学生的档案记录、货物的运输情况”等都是数据。

数据可视化主要旨在借助于图形化手段，清晰有效地传达与沟通信息。为了有效地传达思想概念，美学形式与功能需要齐头并进，通过直观地传达关键的方面与特征，从而实现对于相当稀疏而又复杂的数据集的深入洞察。

数据可视化通过图表直观地展示数据间的量级关系，其目的是将抽象信息转换为具体的图形，将隐藏于数据中的规律直观地展现出来。图表是数据分析可视化最重要的工具，通过点的位置、曲线的走势、图形的面积等形式，直观地呈现研究对象间的数量关系。不同类型的图表展示数据的侧重点不同，选择合适的图表可以更好地进行数据可视化。

1.3 Hadoop生态体系

从当前大数据领域的产业链来看，大数据领域涉及数据采集、数据存储、数据分析和数据应用等环节，不同的环节需要采用不同的技术，且这些环节都要依赖于大数据平台，而Hadoop则是当前比较流行的大数据平台之一。

Hadoop平台经过多年的发展已经形成了一个比较完善的生态体系，而且，由于Hadoop平台是开源的，所以很多商用的大数据平台也是基于Hadoop搭建的，并且形成了极其火爆的技术生态圈，并得到了广泛的应用。

1.3.1 什么是Hadoop

Hadoop是一个开源的用于大规模数据集成的分布式存储和处理的工具平台。它最早由Yahoo!的技术团队根据Google所发布的公开论文思想使用Java语言开发的，现在则隶属于Apache基金会。使用Hadoop平台，用户可以在不了解分布式底层细节的情况下开发分布式程序，充分利用集群的性能进行高速海量数据运算和大数据存储。Hadoop实现了一个分布式文件系统（Hadoop Distributed File System，HDFS）。HDFS具有高容错性的特点，并部署在低价的（Low-cost）硬件上；它提供高吞吐量（High Throughput）来访问应用程序的数据，适合那些有着超大数据集（Large Data Set）的应用程序。HDFS放宽了POSIX的要求，可以以流的形式访问（Streaming Access）文件系统中的数据。

Hadoop框架最核心的设计就是HDFS和MapReduce。HDFS为海量的数据提供了存储功能，而MapReduce为海量的数据提供了计算功能。总之，Hadoop是目前分析海量数据的首选工具，并已被各行各业广泛应用于以下场景。

- 大数据量存储：分布式存储。
- 日志处理。
- 海量计算：并行计算。
- ETL：将数据抽取到Oracle、MySQL、DB2、MongDB及主流数据库。
- 使用HBase进行数据分析、读/写操作——Facebook构建了基于HBase的实时数据分析系统。
- 机器学习：比如ApacheMahout项目（ApacheMahout常见领域有协作筛选、集群、归类）。
- 搜索引擎：Hadoop+Lucene实现。
- 数据挖掘：目前比较流行的广告推荐。

- 通过用户行为特征建模。
- 个性化广告推荐。

1.3.2 Hadoop体系

在大数据领域，Hadoop生态体系不断迭代，已形成一个较大的体系结构，既包含核心组件HDFS、MapReduce和YARN，还包括Hive、Pig、Flume、HBase、Sqoop、Zookeeper、Kafka、Spark等，如图1-3所示。

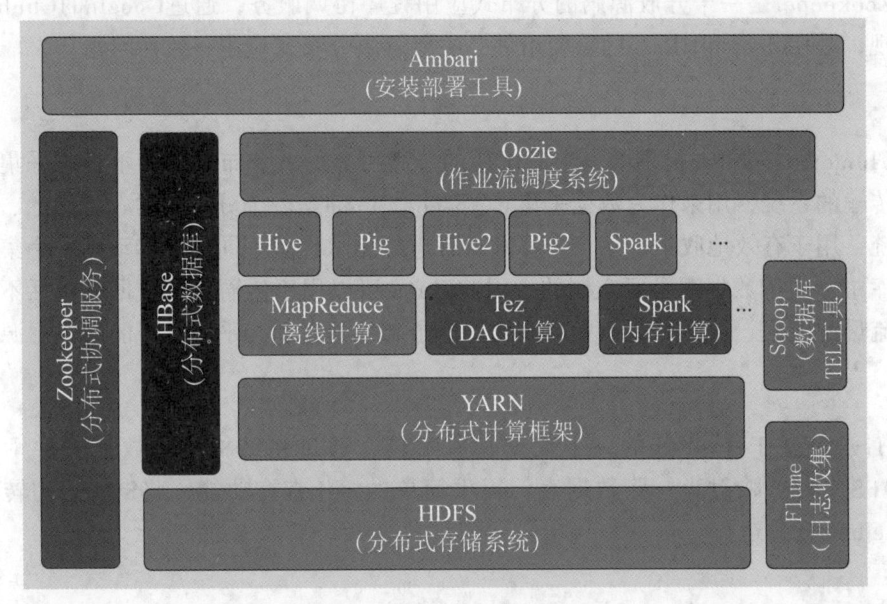

图1-3 Hadoop体系结构

1. HDFS

HDFS是Hadoop中的核心组件之一，具有高容错、高吞吐量等特性，是常用的分布式文件存储系统，它适合一次写入多次读出的场景且不支持文件的修改。因此，HDFS具有能处理超大文件、能流式的访问数据，可以运行于廉价的商用机器集群等。但是该组件具有低延迟数据访问、无法高效存储大量小文件、不支持多用户写入及任意修改文件等缺点。

2. MapReduce

MapReduce是Hadoop中的重要组件之一，简称MR。作为分布式计算模型，开发人员只需在Mapper、Reducer中编写业务逻辑，然后直接交由框架进行分布式计算即可。MapReduce的内核就是分而治之的程序处理理念，把一个复杂的任务划分为若干个简单的任务分别来做。另外，就是程序的调度问题，哪些任务给哪些Mapper来处理是一个要着重考虑的问题。MapReduce的根本原则是信息处理的本地化，哪台PC要处理相应的数据，哪台PC就负责处理该部分的数据，这样做的意义在于可以减少集群网络通信负担。

3. YARN

YARN是Hadoop中的重要组件之一，负责海量数据运算时的资源调度，是在Hadoop 2.X版本中引入的一种新的Hadoop资源管理器，它是一个通用资源管理系统和调度平台，可为上层应用提供统一的资源管理和调度，它的引入为集群在利用率、资源统一管理和数据共享等方面带来了巨大好处。

4. Zookeeper

Zookeeper是一个开放源码的分布式应用程序协调服务，通过Google的Chubby开源实现，是Hadoop和HBase的重要组件，主要解决分布式应用一致性问题。

5. Flume

Flume是Cloudera提供的一个高可用的、高可靠的、分布式的海量日志采集、聚合和传输的系统，用来作为数据采集。Apache Flume是一个分布式的、可靠的、可用的系统，用于有效地收集、聚合和将大量日志数据从许多不同的源移动到一个集中的数据库中。由于数据源是可定制的，因此Flume可以用于传输大量数据，包括不限于网络流量的数据、社交媒体生成的数据、电子邮件消息和几乎所有可能的数据源。

6. Hive

Hive是基于Hadoop的一个数据仓库工具，通过将结构化的数据文件（通常为HDFS文件）映射为一张数据表，提供简单的SQL查询功能，将SQL语句转换为MapReduce任务运行。

7. HBase

HBase是一个建立在Hadoop文件系统之上的面向列的分布式数据库。不同于一般的关系数据库，HBase适合存储非结构化的数据，HBase基于列而不是基于行。

8. Sqoop

主要用于在Hadoop(Hive)与传统的数据库(MySQL、PostgreSQL)间进行数据的传递，可以将关系型数据库（例如MySQL、Oracle、Postgres等）中的数据导入Hadoop的HDFS中，也可以将HDFS中的数据导入关系型数据库中。

9. Kafka

Kafka是一个高吞吐量的分布式消息订阅-发布系统，通过与Spark Streaming整合，完成实时业务计算，使用Java+Scala开发。Kafka的目的是通过Hadoop的并行加载机制来统一处理线上和离线的消息，也是为了通过集群来提供实时的消息。

10. Spark

Spark是大规模数据快速处理通用的计算引擎，其提供大量的库，即Spark Core、Spark SQL、Spark Streaming、MLlib、GraphX（只进行计算，不进行存储）。Spark专门用于大数据量下的迭代计算。Spark是为了与Hadoop配合而开发出来的，不是为了取代Hadoop。Spark的运算速度比Hadoop的MapReduce框架的运算速度快的原因是，Hadoop在第一次MapReduce运算之后，会将数据的运算结果从内存写入磁盘中，第二次MapReduce运算时再从磁盘中读取数据，所以其瓶颈是两次运算间的多余I/O消耗。Spark则是将数据一直缓存在内存中，直到得到最后的结果，再将结果写入磁盘，所以多次运算的情况下，Spark的运算速度是比较快的。

1.4 本章小结

本章首先介绍了大数据的发展历程、特征、思维方式，随后重点介绍了大数据的开发流程，最后介绍了Hadoop生态体系的重要组件以及应用场景。

“十四五”时期是我国工业经济向数字经济迈进的关键时期，对大数据产业的发展提出了新的要求。大数据是人类社会生产活动信息化的产物，不仅包含海量数据的存储，还包括其相关技术、领域应用等内容，而Hadoop是大数据技术的标准，具有广泛的应用。本章介绍虽然简单，但可以为后面各章Hadoop相关技术的深入学习打下基础。

1.5 课后习题

1.简答题

（1）结合你身边的大数据项目，针对大数据应用开发流程说说你的认识。

（2）“十四五”规划对大数据行业带来了什么要求。

（3）Hadoop就是大数据是否正确，说出你的观点。

第2章 Hadoop平台部署

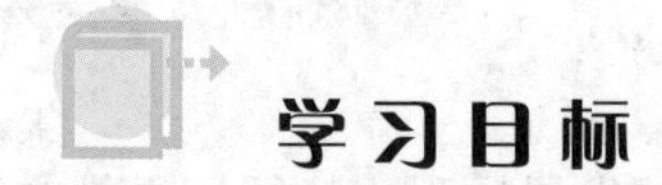

学习目标

（1）掌握虚拟机的创建方法和CentOS 7操作系统的安装方法。

（2）掌握CentOS 7虚拟机的基本配置方法，包括修改主机名、设置固定IP地址、关闭防火墙和新建安装目录等。

（3）掌握为CentOS虚拟机安装和配置JDK的方法。

（4）掌握CentOS虚拟机的克隆方法和主机IP地址映射的配置方法。

（5）掌握集群各节点SSH免密码登录的配置方法。

（6）掌握Hadoop三种模式的安装、配置、启动与测试方法。

思政目标

（1）了解我国在大数据方面的基础软件项目。

（2）理解服务器安全对国家安全的重要性。

Hadoop集群可以安装在各个操作系统。在实际应用中，CentOS 7是使用广泛的操作系统，本文选择在CentOS 7上搭建Hadoop集群。为了方便教学，本节将在VMware Workstation软件中创建虚拟机，安装CentOS 7系统，并搭建集群。

2.1 安装准备

虚拟机是指通过软件模拟的具有完整硬件系统功能的、运行在一个完全隔离环境中的完整计算机系统。常用的虚拟机有VMware Workstation、Virtual PC、VirtualBox，本节采用VMware Workstation 12 PRO。安装成功后，操作界面如图2-1所示。

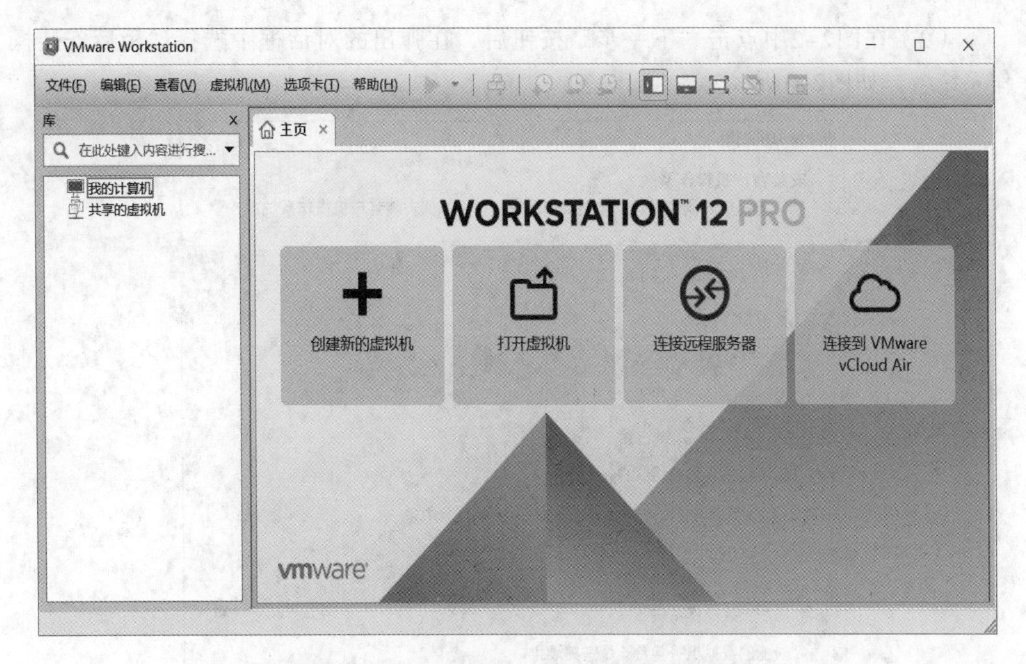

图2-1 VMware Workstation 12 PRO操作界面

2.1.1 虚拟机安装

微课：v2-1 创建虚拟机

新建虚拟机的操作步骤如下。

（1）在图2-1中，单击“创建新的虚拟机”选项，选择“典型（推荐）”，如图2-2所示。

图2-2 选择安装类型

（2）在图2-2中点击“下一步”按钮后，在弹出的对话框中选择“稍后安装操作系统”，如图2-3所示。

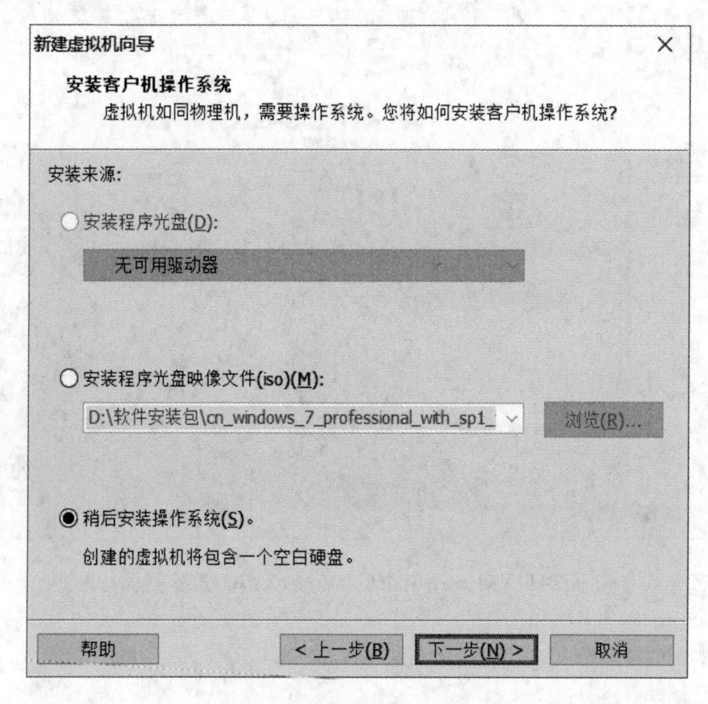

图2-3 安装操作系统选项1

（3）在图2-3中点击“下一步”按钮后，在弹出的对话框中选择“Linux”、“CentOS 64位”，如图2-4所示。

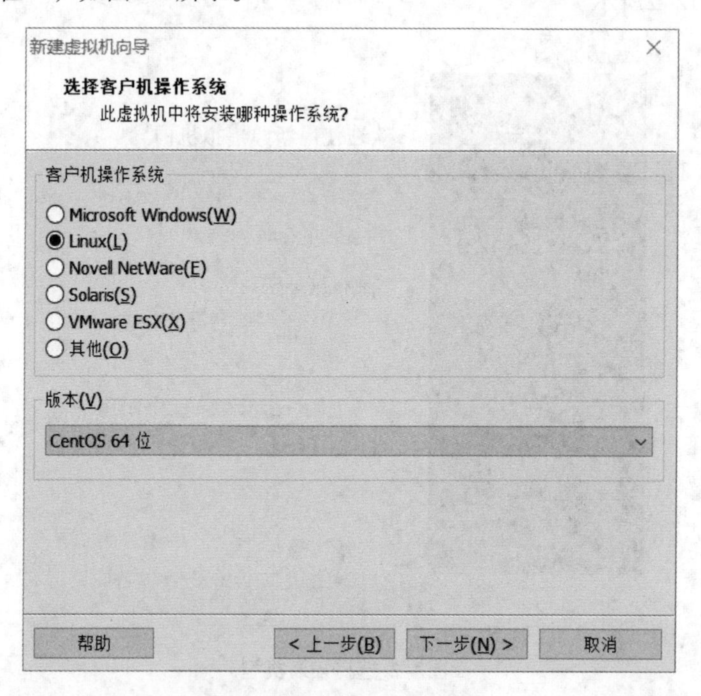

图2-4 安装操作系统选项2

（4）在图2-4中点击“下一步”按钮后，在弹出的对话框中设置虚拟机名称，并配置安装目录，如图2-5所示。

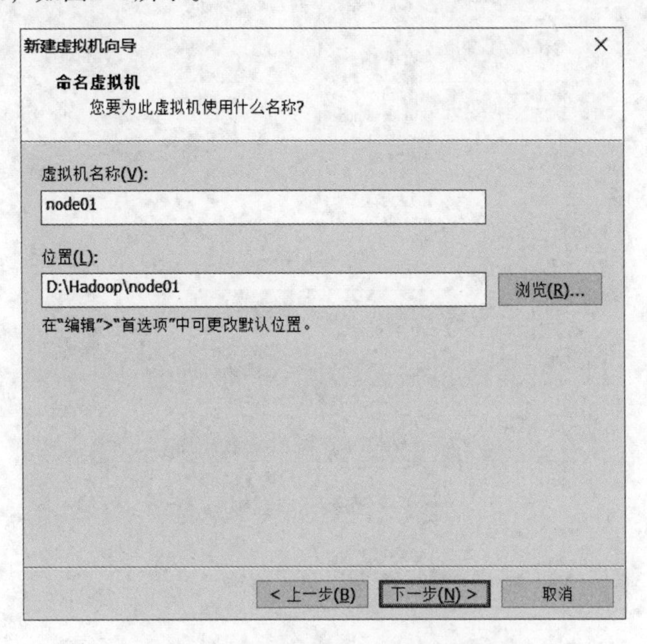

图2-5 设置虚拟机名称并配置安装目录

（5）在图2-5中点击“下一步”按钮后，在弹出的对话框中点击“将虚拟磁盘拆分成多个文件”，如图2-6所示。

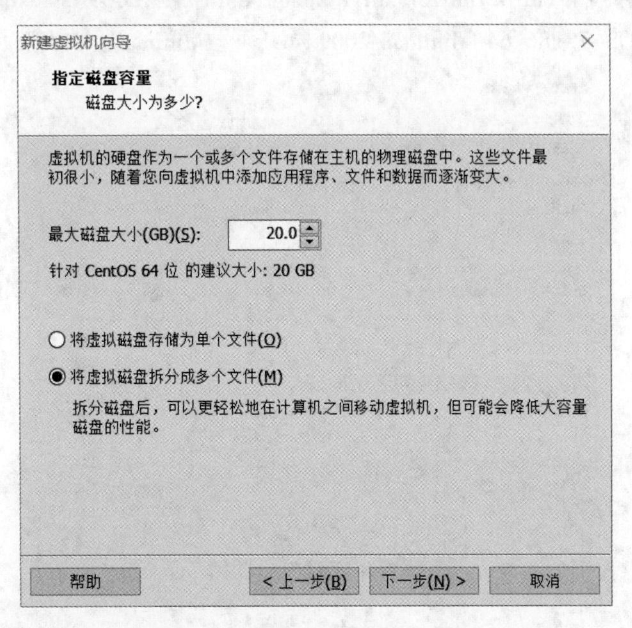

图2-6 将虚拟机拆分成多个文件

（6）在图2-6中点击“下一步”按钮后，预览虚拟机，点击“完成”按钮，完成虚拟机的创建，如图2-7所示。

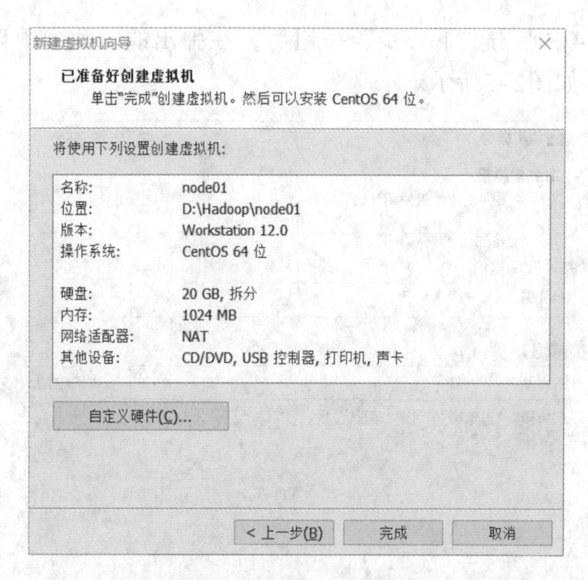

图2-7 安装完成虚拟机

2.1.2 安装CentOS 7操作系统

微课：v2-2 安装CentOS 7操作系统

1. 下载CentOS 7操作系统

刚创建好的虚拟机可以认为是一个虚拟出的只有硬件没有操作系统的设备，如果要让虚拟机运行，则需要为虚拟机安装操作系统。

打开浏览器，输入http://mirrors.aliyun.com/centos/7.9.2009/isos/x86_64/，下载操作系统，单击CentOS-7-x86_64-Minimal-2009.iso，下载Minimal版本安装，如图2-8所示。

阿里云开源镜像站资源目录

阿里云 开源镜像站

Index of /centos/7.9.2009/isos/x86_64/

File Name	File Size	Date
Parent directory/	-	-
0_README.txt	2.4 KB	2020-11-06 10:32
CentOS-7-x86_64-DVD-2009.iso	4.4 GB	2020-11-04 07:37
CentOS-7-x86_64-DVD-2009.torrent	176.1 KB	2020-11-06 10:44
CentOS-7-x86_64-Everything-2009.iso	9.5 GB	2020-11-02 11:18
CentOS-7-x86_64-Everything-2009.torrent	380.6 KB	2020-11-06 10:44
CentOS-7-x86_64-Minimal-2009.iso	973.0 MB	2020-11-03 10:55
CentOS-7-x86_64-Minimal-2009.torrent	38.6 KB	2020-11-06 10:44
CentOS-7-x86_64-NetInstall-2009.iso	575.0 MB	2020-10-27 12:26
CentOS-7-x86_64-NetInstall-2009.torrent	23.0 KB	2020-11-06 10:44
sha256sum.txt	398.0 B	2020-11-04 07:38
sha256sum.txt.asc	1.2 KB	2020-11-06 10:37

图2-8 下载CentOS 7

提 示

由于CentOS官方网站服务器架设在法国，网络波动较大，下载时容易失败，因此可以选择国内镜像站点下载，如表2-1所示。

表2-1 国内CentOS镜像

http://mirrors.bupt.edu.cn/centos/7.9.2009/isos/x86_64/	北京邮电大学
http://mirrors.njupt.edu.cn/centos/7.9.2009/isos/x86_64/	南京邮电大学
http://mirrors.bfsu.edu.cn/centos/7.9.2009/isos/x86_64/	北京外国语大学
http://ftp.sjtu.edu.cn/centos/7.9.2009/isos/x86_64/	上海交通大学
http://mirrors.cqu.edu.cn/CentOS/7.9.2009/isos/x86_64/	重庆大学
http://mirrors.aliyun.com/centos/7.9.2009/isos/x86_64/	阿里云
http://mirrors.nju.edu.cn/centos/7.9.2009/isos/x86_64/	南京大学
http://mirrors.tuna.tsinghua.edu.cn/centos/7.9.2009/isos/x86_64	清华大学
http://mirrors.ustc.edu.cn/centos/7.9.2009/isos/x86_64/	中国科学技术大学
http://mirrors.163.com/centos/7.9.2009/isos/x86_64/	网易
http://mirror.lzu.edu.cn/centos/7.9.2009/isos/x86_64/	兰州大学

2. 载入镜像到虚拟机

（1）在VMware界面，点击选中已经创建好的虚拟机“node01”，可以看到该虚拟机处于关机状态，如图2-9所示。

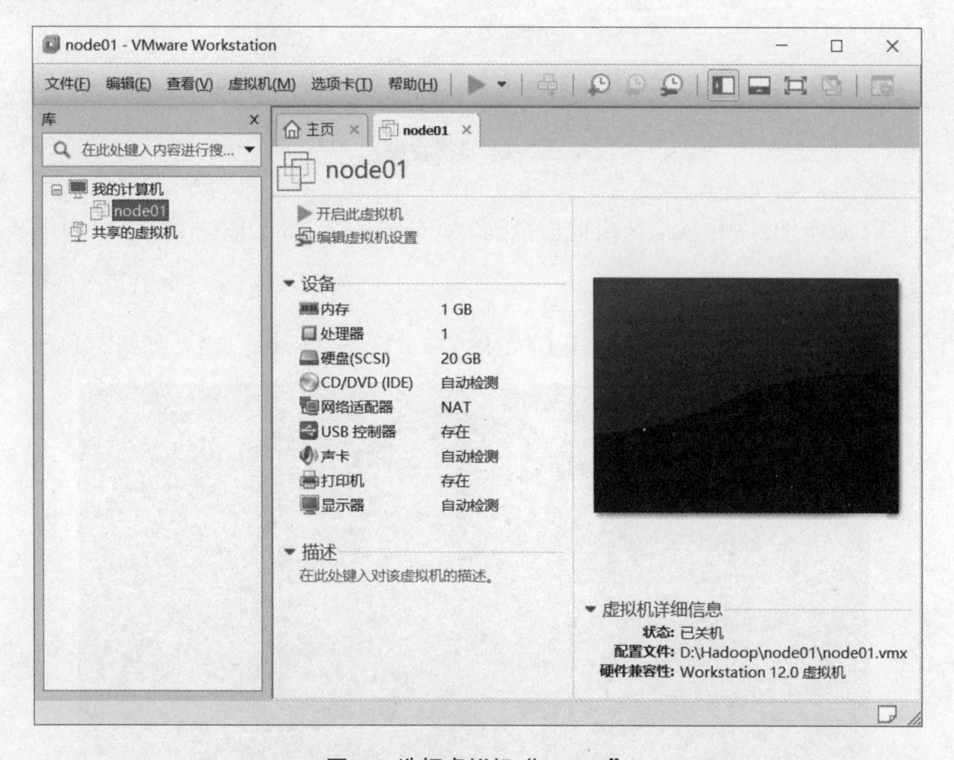

图2-9 选择虚拟机“node01”

（2）在图2-9所示界面中点击“CD/DVD（IDE）”，在弹出的界面中选择“使用ISO映像文件”，并浏览选择下载的镜像，点击“确认”按钮即可完成镜像选择，如图2-10所示。

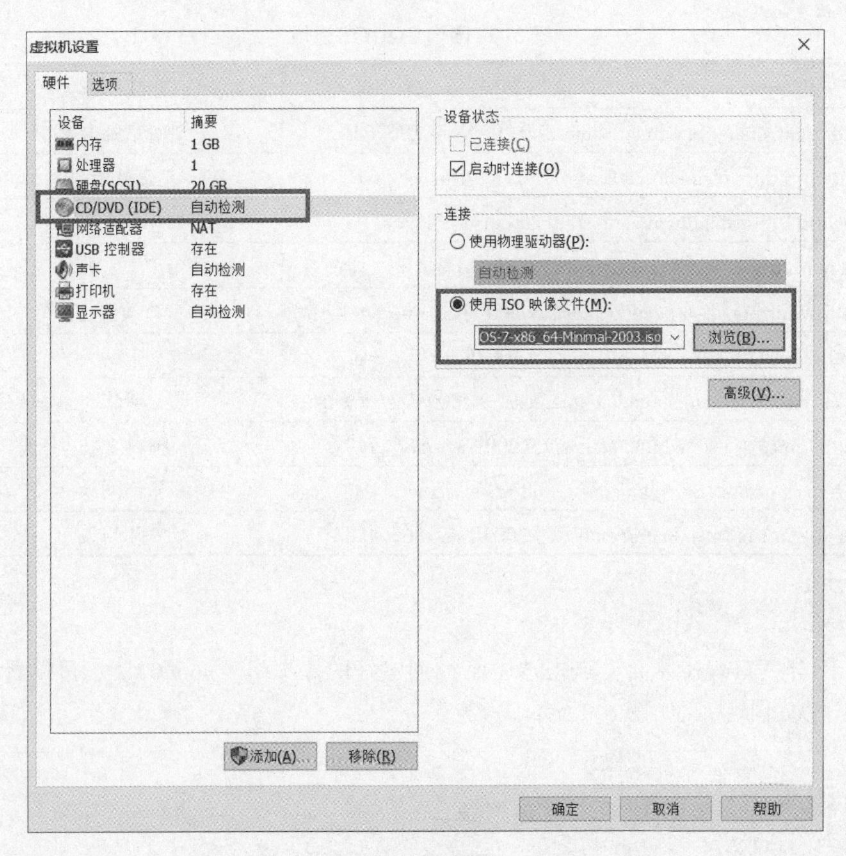

图2-10 选择镜像

（3）点击图2-9中的“开启此虚拟机”，开始安装CentOS 7系统。

3. CentOS 7安装

（1）通过方向键选择“Install CentOS 7”，按下“Enter”键，如图2-11所示。

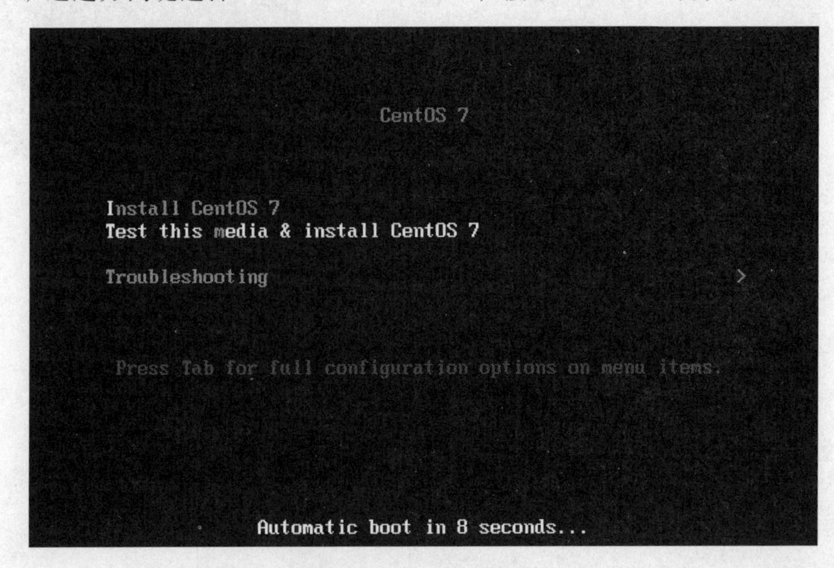

图2-11 选择安装选项

（2）选择系统语言。选择“English”作为安装过程中的语言，点击“Continue”按钮继续安装，如图2-12所示。

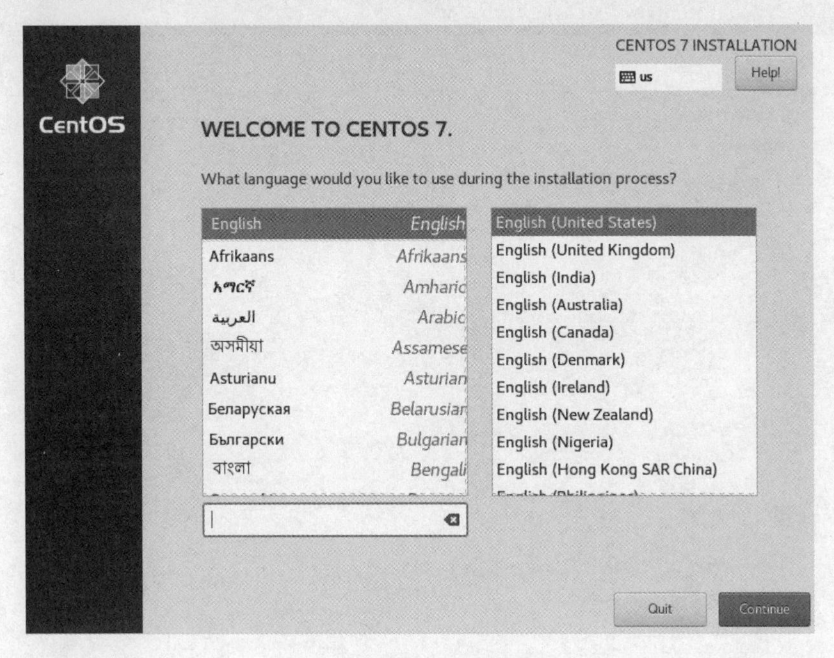

图2-12 选择系统安装语言

（3）在安装界面找到“SYSTEM”大类，点击“INSTALLATION DESTINATION”，如图2-13所示，进入系统磁盘配置选项，再点击“Done”按钮完成操作，如图2-14所示，在返回的界面中选择“Begin Installation”按钮开始安装系统，如图2-15所示。

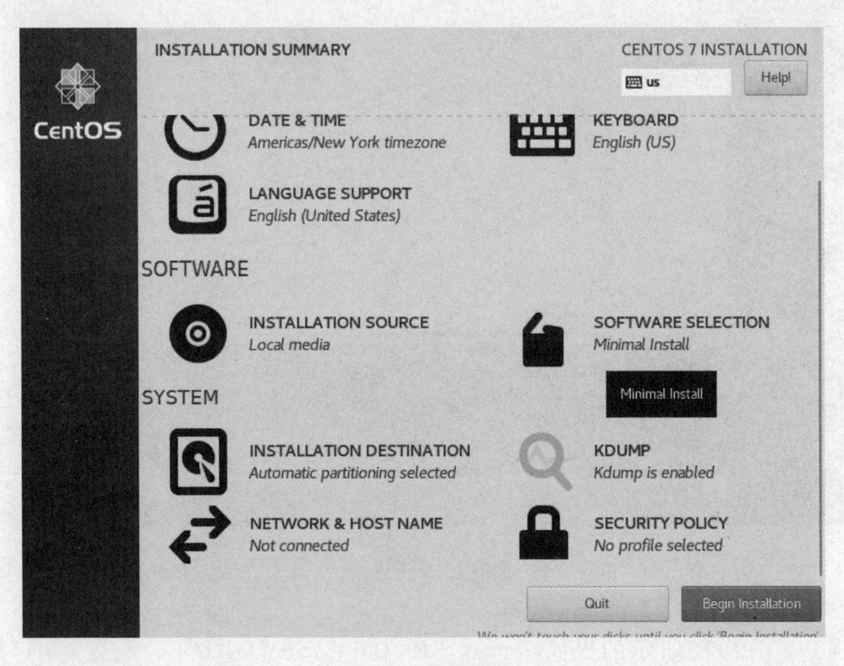

图2-13 选择安装

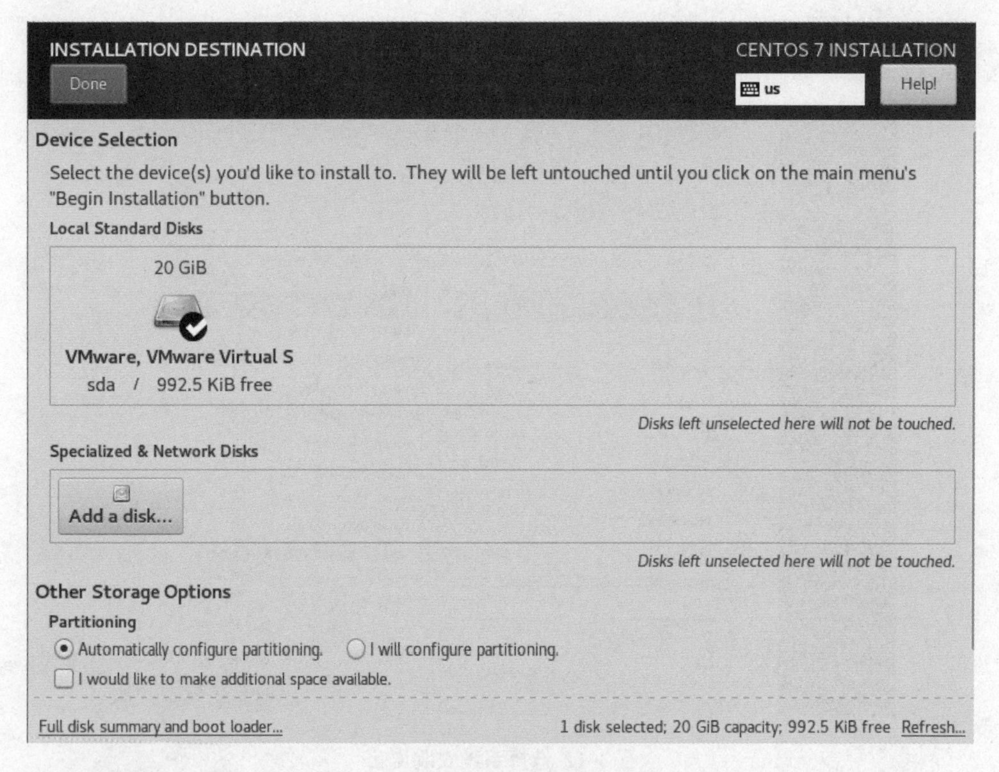

图2–14 完成设置

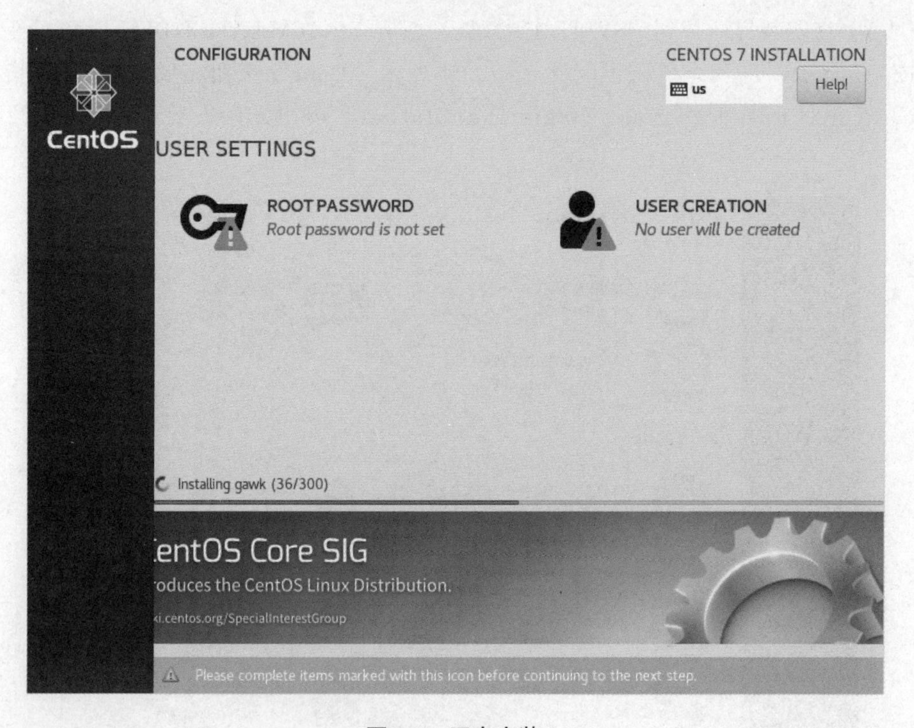

图2–15 正在安装

（4）在图2–15所示的界面中选择“ROOT PASSWORD”，配置ROOT用户密码（要牢记，学习阶段建议设为123456），并重复输入验证密码，如图2–16所示。

完成后点击“Done”按钮完成密码设置，等待安装完成。当系统完成安装后，点击“Reboot”按钮重启系统，完成CentOS 7安装，如图2-17所示。

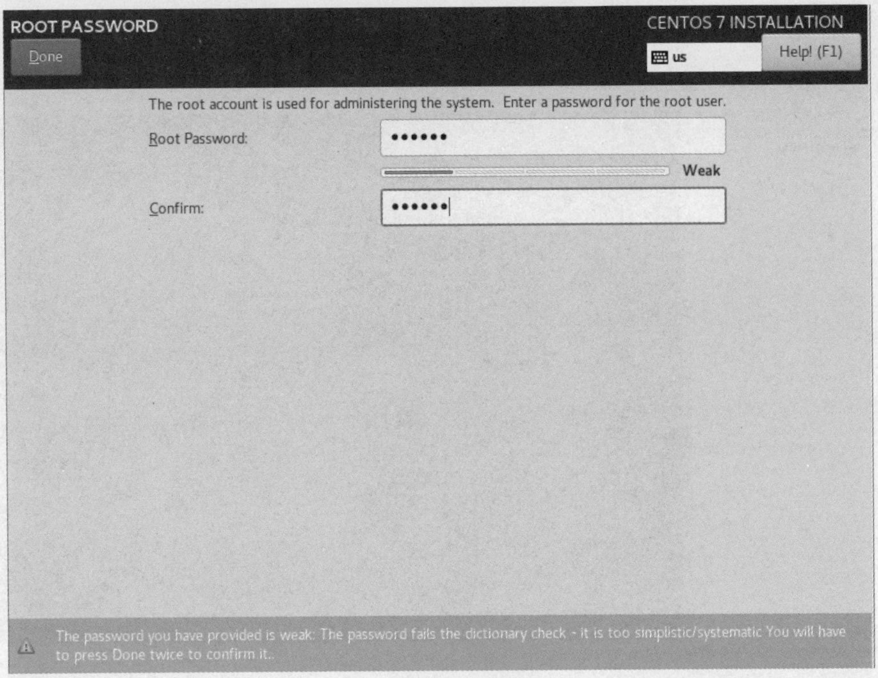

图2-16 设置密码

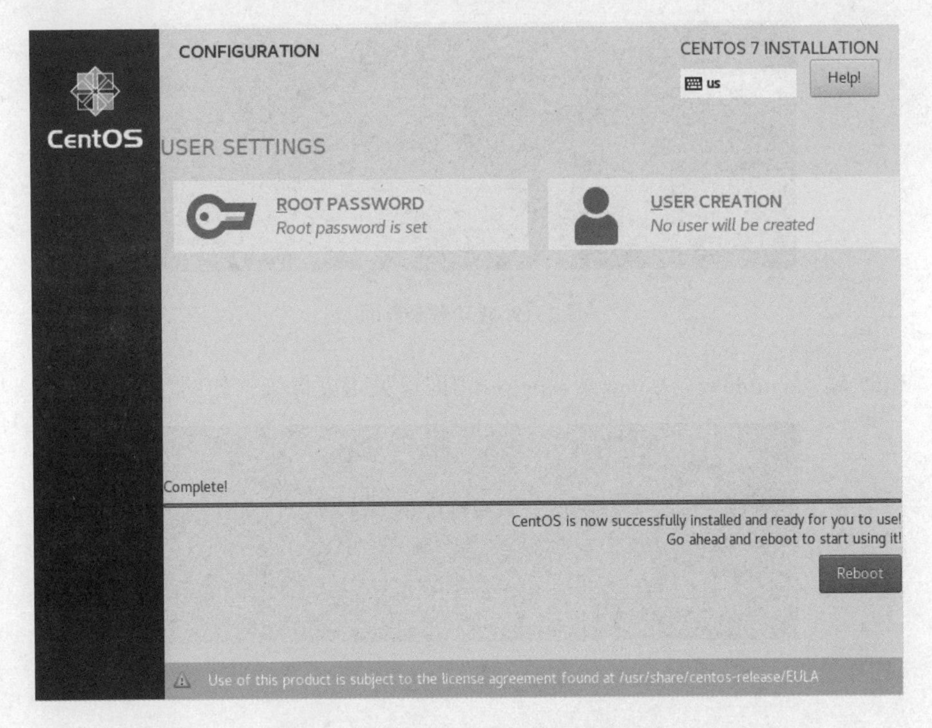

图2-17 完成CentOS 7安装并重启系统

（5）完成CentOS 7安装步骤后，系统重新启动，在登录界面输入用户名“root”，按“Enter”键确认，随后输入密码“123456”，按“Enter”键确认登录，如图2-18所示，登录界面如图2-19所示。

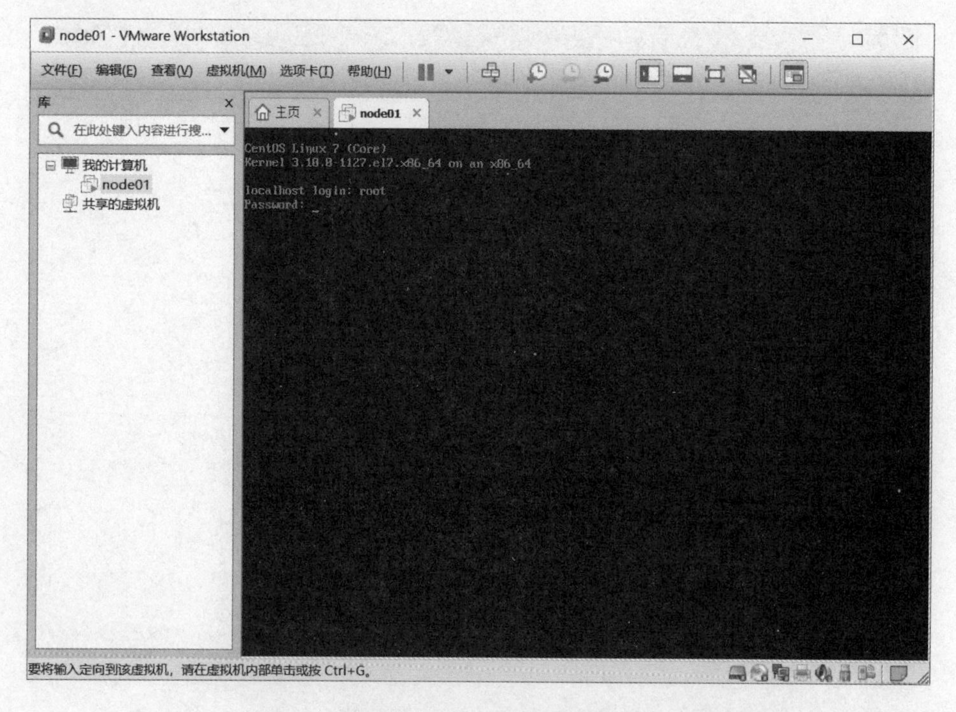

图2-18 CentOS 7启动界面

```
CentOS Linux 7 (Core)
Kernel 3.10.0-1127.el7.x86_64 on an x86_64

localhost login: root
Password:
Last login: Thu Feb 24 04:36:31 on tty1
[root@localhost ~]#
```

图2-19 成功登录界面

（6）输入shutdown -h now或者poweroff即可关闭虚拟机，如图2-20所示。

```
CentOS Linux 7 (Core)
Kernel 3.10.0-1127.el7.x86_64 on an x86_64

localhost login: root
Password:
Last login: Thu Feb 24 04:42:45 on tty1
[root@localhost ~]# shutdown -h now _
```

图2-20 关机虚拟机

4. 虚拟机备份

初学者在操作CentOS 7系统时，容易出现误操作导致系统文件受损、误删数据等情况，VMware的快照功能可以在出现上述问题时恢复系统。

磁盘“快照”是虚拟机磁盘文件（VMDK）在某个时间点的副本。系统崩溃或系统异常，你可以通过快照来还原磁盘文件系统。VMware快照是VMware Workstation里的一个特色功能。

（1）创建快照。

在虚拟机关闭状态，选中虚拟机，点击右键，选择“快照”→“拍摄快照”，如图2-21所示。

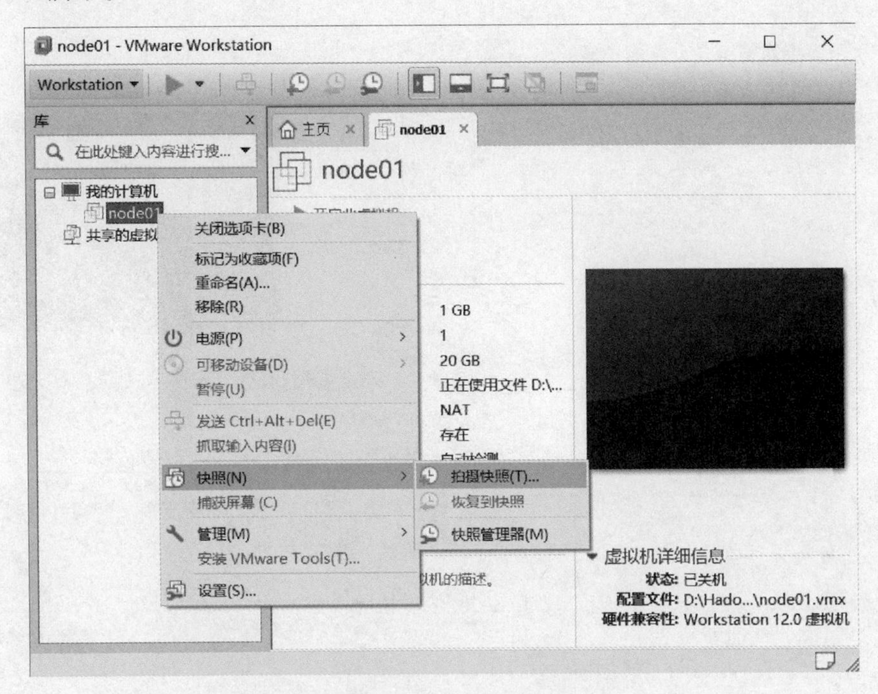

图2-21 选择“快照”→“拍摄快照”

（2）设置快照名称。

当快照拍摄完成后，需要设置快照名称，本节将快照名称设为“初始的centos7”，用来保存初始的centos7状态，如图2-22所示。

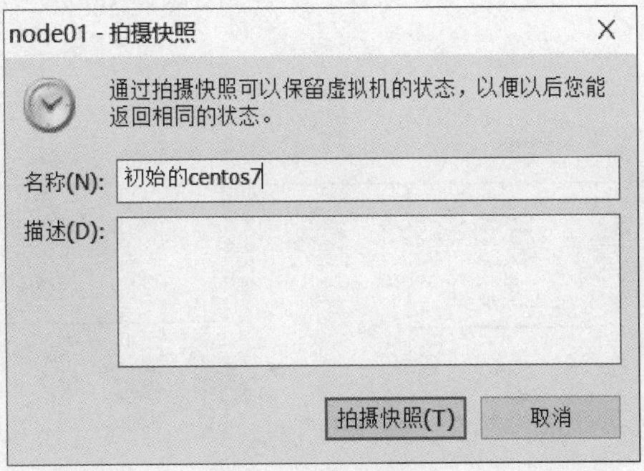

node01 - 拍摄快照

通过拍摄快照可以保留虚拟机的状态，以便以后您能返回相同的状态。

名称(N): 初始的centos7

描述(D):

拍摄快照(T) 取消

图2-22 初始的快照

（3）恢复快照。

当虚拟机需要恢复拍摄快照时，可以选中虚拟机，右击选中“快照”，在弹出的菜单中选择“恢复到快照（R）:初始的centos7”，如图2-23所示。

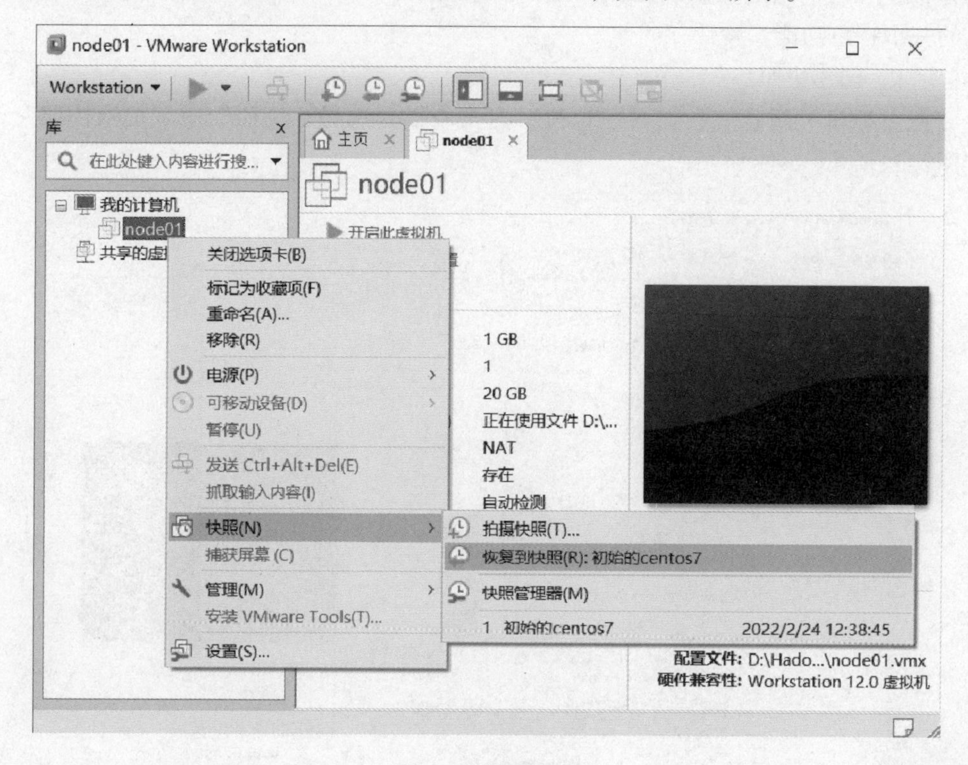

图2-23 恢复快照

2.1.3 CentOS 7常用指令

微课：
v2-3
CentOS 7常用指令

在搭建Hadoop环境时，Linux基本命令的用法是必须掌握的，它们与目录、文件、网络等操作息息相关。

1. 目录操作

Linux的目录结构为树状结构，顶级目录为根目录/。CentOS 7树状目录结构如图2-24所示。

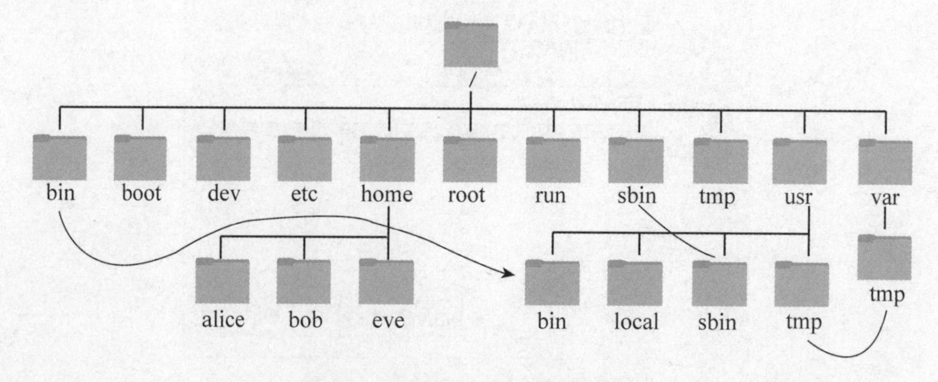

图2-24 CentOS 7树状目录结构

CentOS 7目录操作常用命令如表2-2所示。

表2-2 CentOS 7目录操作常用命令

命令	英文含义	中文释义
ls	list files	列出目录及文件名
cd	change directory	切换目录
pwd	print work directory	显示目前所在的目录
mkdir	make directory	创建新目录
cp	copy file	复制文件或目录
rm	remove	删除文件或目录
mv	move file	移动文件与目录或修改文件与目录的名称

（1）ls（列出目录及文件名）。

ls可能是最常被运行的命令。单独输入ls，为展示当前目录及文件名列表；输入ls -l，为展示当前详细目录与文件列表；当ls后附加指定目录时，则为查看指定目录内的目录及文件名列表。操作如下：

```
[root@localhost ~]# ls
anaconda-ks.cfg
[root@localhost ~]# ls -l
total 4
-rw-------.1 root root 1272 Feb 23 17:18 anaconda-ks.cfg
[root@localhost ~]# ls /
bin  boot  dev  etc  home  lib  lib64  media  mnt  opt  proc  root  run
sbin srv   sys  tmp  usr   var
```

（2）cd（切换目录）。

cd可切换当前目录，默认目录为home，可以使用绝对路径或相对路径。直接输入cd命令，将切换到你的home目录下，不管你当前所在的目录是什么。操作如下：

```
[root@localhost/]# cd
[root@localhost ~]#
```

cd+可以切换至指定目录，操作如下：

```
[root@localhost ~]# cd /usr/local/
[root@localhost local]#
```

cd..可以返回上一级目录，操作如下：

```
[root@localhost usr]# cd /usr/local/
[root@localhost local]# cd..
[root@localhost usr]#
```

（3）pwd(显示当前所在的目录)。

pwd可以显示当前所在的目录，操作如下：

```
[root@localhost usr]# cd /usr/local/
[root@localhost local]# pwd
/usr/local
```

（4）mkdir(创建新目录)。

mkdir可以创建新目录，操作如下：

```
[root@localhost ~]# mkdir hadoop
[root@localhost ~]# ls
anaconda-ks.cfg hadoop
```

（5）cp(复制文件或目录)。

cp可以复制文件或目录，操作如下：

```
[root@localhost ~]# ls
anaconda-ks.cfg
[root@localhost ~]# cp anaconda-ks.cfg 1.txt #将anaconda-ks.cfg复制为1.txt
[root@localhost ~]# ls
1.txt anaconda-ks.cfg
```

（6）rm(删除文件或目录)。

rm可以删除文件或目录，操作如下：

```
[root@localhost ~]# ls
1.txt    hadoop                          #列表中有文件1.txt和文件夹hadoop
[root@localhost ~]# rm 1.txt             #删除文件时只需要rm命令
rm:remove regular file '1.txt' ? yes     #确认删除
[root@localhost ~]# rm -r hadoop/        #删除文件夹时需要参数-r
rm:remove directory 'hadoop/' ? yes      #确认删除
```

（7）mv(移动文件与目录或修改文件与目录的名称)。

mv可以移动文件与目录或修改文件与目录的名称，操作如下：

```
[root@localhost ~]# ls
anaconda-ks.cfg
[root@localhost ~]# mv anaconda-ks.cfg 1.txt #将anaconda-ks.cfg修改为1.txt
[root@localhost ~]# ls
1.txt
```

2. vi操作

vi是Linux系统下标准的编辑器，它的强大不逊色于任何最新的文本编辑器，这里只是简单地介绍它的用法和一小部分指令。由于对Linux系统的任何版本，vi编辑器是完全相同的，因此你可以在其他任何介绍vi的地方进一步了解它。vi也是Linux中最基本的文本编辑器，学会它后，你将在Linux的世界里畅行无阻。

（1）启动vi，操作如下：

```
[root@localhost ~]# vi abc.txt    //abc.txt可以是路径下不存在的文件名
```

（2）vi模式转换。

vi有指令模式(Command Mode)和输入模式(Insert Mode)两种。进入vi界面后，默认进入指令模式，或者在vi界面按“Esc”键，也可以进入命令模式。从命令模式输入i、o等字符也可以进入输入模式。其中，i代表从光标所在处前面一个字母开始输入，o代表新开一行并进入输入模式。

（3）存储和退出。

在命令模式输入“:wq”，代表存储并退出。“:w”代表存储文件。当没有修改文件内容时，可以输入“:q”退出vi。当输入“:q!”时，表示不保存文件并退出。

3. 系统状态

（1）查看当前主机名的命令如下：

```
[root@localhost ~]# hostname
```

（2）查看系统时间的命令如下：

```
[root@localhost ~]# date
```

（3）查看进程状态的命令如下：

```
[root@localhost ~]# ps -aux
```

（4）结束正在进行的指定进程，通过上一行命令可以查询正在活动的进程PID，使用kill命令可以关闭活动的指定进程，如下：

```
[root@localhost ~]# kill -9活动的进程PID
```

（5）磁盘分区信息的命令如下：

```
[root@localhost ~]# df -h
```

2.1.4 网络配置

微课：
v2-4
CentOS 7网络配置

1. 进入网络配置文件目录

命令如下，ifcfg-ens33就是需要配置的网络文件，如图2-25所示。

```
[root@localhost ~]# cd /etc/sysconfig/network-scripts
```

```
[root@localhost ~]# cd /etc/sysconfig/network-scripts
[root@localhost network-scripts]# ls
ifcfg-ens33  ifdown-isdn       ifdown-tunnel  ifup-isdn    ifup-Team
ifcfg-lo     ifdown-post       ifup           ifup-plip    ifup-TeamPort
ifdown       ifdown-ppp        ifup-aliases   ifup-plusb   ifup-tunnel
ifdown-bnep  ifdown-routes     ifup-bnep      ifup-post    ifup-wireless
ifdown-eth   ifdown-sit        ifup-eth       ifup-ppp     init.ipv6-global
ifdown-ippp  ifdown-Team       ifup-ippp      ifup-routes  network-functions
ifdown-ipv6  ifdown-TeamPort   ifup-ipv6      ifup-sit     network-functions-ipv6
```

图2-25 网络配置文件1

2. 动态获取IP地址

使用vi编辑器打开并修改ifcfg-ens33文件，命令如下所示：

```
[root@localhost network-scripts]# vi ifcfg-ens33
```

或者使用绝对路径打开网络配置文件，命令如下所示：

```
[root@localhost ~]# vi /etc/sysconfig/network-scripts/ifcfg-ens33
```

执行上述命令，打开CentOS的网络配置文件，如图2-26所示。

```
TYPE=Ethernet
PROXY_METHOD=none
BROWSER_ONLY=no
BOOTPROTO=dhcp
DEFROUTE=yes
IPV4_FAILURE_FATAL=no
IPV6INIT=yes
IPV6_AUTOCONF=yes
IPV6_DEFROUTE=yes
IPV6_FAILURE_FATAL=no
IPV6_ADDR_GEN_MODE=stable-privacy
NAME=ens33
UUID=a5211f41-4f33-4f65-8cd6-ebb1e677210d
DEVICE=ens33
ONBOOT=no
```

图2-26 网络配置文件2

在图2-26所示的配置界面中，ONBOOT=no表示网卡默认未开启，需要修改为ONBOOT=yes，保存并退出文件。再使用如下命令重启网络服务，即可实现动态IP地址，如下：

```
[root@localhost ~]# service network restart
```

使用如下命令可以查看动态获取的IP地址，如图2-27所示。

```
[root@localhost ~]# ip addr
```

```
[root@localhost ~]# ip addr
1: lo: <LOOPBACK,UP,LOWER_UP> mtu 65536 qdisc noqueue state UNKNOWN group default qlen 100
0
    link/loopback 00:00:00:00:00:00 brd 00:00:00:00:00:00
    inet 127.0.0.1/8 scope host lo
       valid_lft forever preferred_lft forever
    inet6 ::1/128 scope host
       valid_lft forever preferred_lft forever
2: ens33: <BROADCAST,MULTICAST,UP,LOWER_UP> mtu 1500 qdisc pfifo_fast state UP group defau
lt qlen 1000
    link/ether 00:0c:29:9d:57:74 brd ff:ff:ff:ff:ff:ff
    inet 192.168.137.140 24 brd 192.168.137.255 scope global noprefixroute dynamic ens33
       valid_lft 975sec preferred_lft 975sec
    inet6 fe80::2a51:d141:258e:ac23/64 scope link noprefixroute
       valid_lft forever preferred_lft forever
```

图2-27 查看动态获取的IP地址

3. 设置静态IP地址

当需要搭建Hadoop集群时，由于有多台设备参与运算，所以需要分别为设备设置静态IP地址，以便管理。

首先需要查看VMware虚拟网卡设定的IP地址范围，在VMware主界面菜单栏点击“编辑”，选择“虚拟网络编辑器”，在弹出的对话框中选择“VMnet8”选项，可查看VMware软件划分的子网IP段，如图2-28所示。

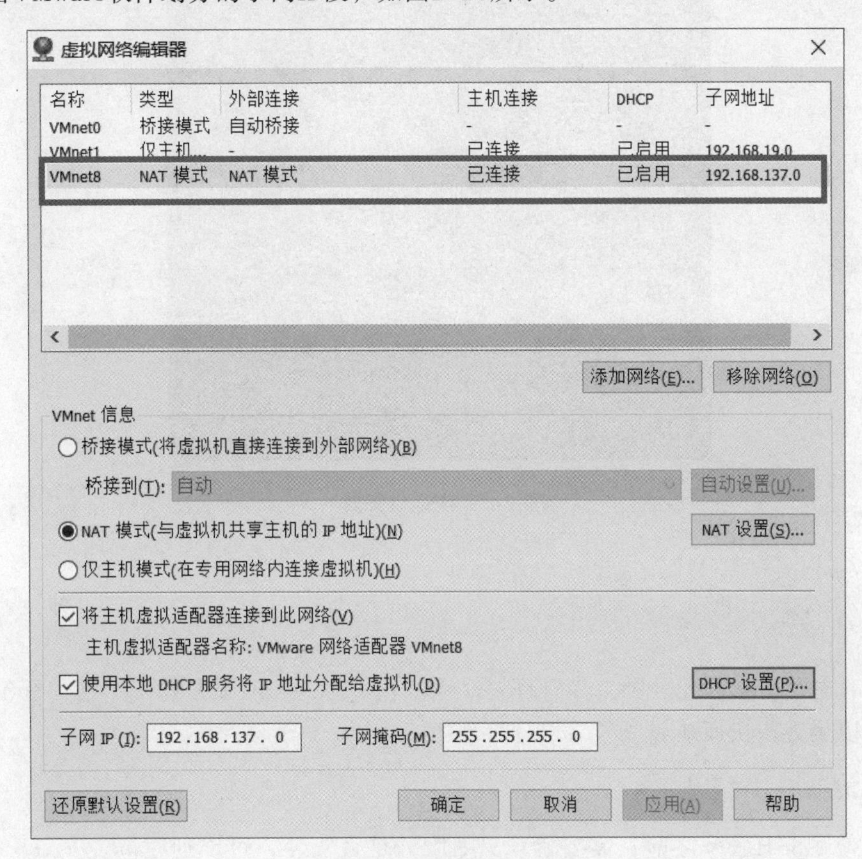

图2-28 查看VMnet8子网

图2-28中划分的子网地址为192.168.137.0，因此，CentOS 7可设置的静态IP地址范围为192.168.137.0~192.168.137.255。在配置时，建议IP地址最后一位选择范围在100~255之间。每一台主机划分的子网IP地址都可能不一样，读者需根据自己的实际情况去划分静态IP地址。

配置静态IP地址的操作与自动获取IP地址的操作相同。使用如下命令打开网络设置：

```
[root@localhost ~]# vi /etc/sysconfig/network-scripts/ifcfg-ens33
```

IP地址配置如图2-29所示。

在图2-29中：BOOTPROTO=static表示由原有的DHCP（动态获取IP地址）切换为static（静态获取IP地址）。ONBOOT=yes表示启动该网卡。IPADDR表示设定IP地址，该地址由观察VMnet8虚拟网卡子网地址设定。GATEWAY表示虚拟网关，一般由VMnet8虚拟网卡子网地址设定，最后一位数设定为2。NETMASK表示虚拟机子网掩码，设定为255.255.255.0。DNS1表示域名解析器，本书使用国内通用的DNS服务器114.114.114.114，也可自行选择其他DNS服务器，或者自行搭建DNS服务器。

最后需要重启网络服务，查看配置好的IP地址。

```
TYPE=Ethernet
PROXY_METHOD=none
BROWSER_ONLY=no
BOOTPROTO=static
DEFROUTE=yes
IPV4_FAILURE_FATAL=no
IPV6INIT=yes
IPV6_AUTOCONF=yes
IPV6_DEFROUTE=yes
IPV6_FAILURE_FATAL=no
IPV6_ADDR_GEN_MODE=stable-privacy
NAME=ens33
UUID=a5211f41-4f33-4f65-8cd6-ebb1e677210d
DEVICE=ens33
ONBOOT=yes
IPADDR=192.168.137.100
GATEWAY=192.168.137.2
NETMASK=255.255.255.0
DNS1=114.114.114.114
```

图2-29 IP地址配置

2.1.5 SSH服务配置

SSH主要用于远程登录。当管理多个节点时，通过SSH服务并使用远程登录管理工具可以更方便快捷地对服务器进行管理。

1. 配置SSH远程服务

在使用SSH服务之前，需要确认SSH服务是否安装，可以通过ps -e|grep ssh验证，如图2-30所示。

```
[root@localhost ~]# ps -e|grep ssh
  1025 ?        00:00:00 sshd
  1280 ?        00:00:00 sshd
  1799 ?        00:00:00 sshd
```

图2-30 查看SSH服务是否安装和开启

如图2-30所示，CentOS虚拟机已经默认安装并开启了SSH服务，不需要进行额外安装。如果系统未安装该服务，则可以通过如下指令安装：

```
[root@localhost ~]# yum install openssh-server
```

2. 远程登录工具

微课：v2-5 远程连接

常用的SSH远程登录工具有PuTTY、Xshell、MobaXterm、SecureCRT、FinalShell。本书采用FinalShell。读者可以登录http://www.hostbuf.com下载最新版本。

FinalShell是一款国产免费远程登录工具。它不仅是一体化的SSH客户端，还是功能强大的开发运维工具，能充分满足开发运维需求。同时，它还支持SFTP协议，可以实现文件传递。FinalShell主界面如图2-31所示。

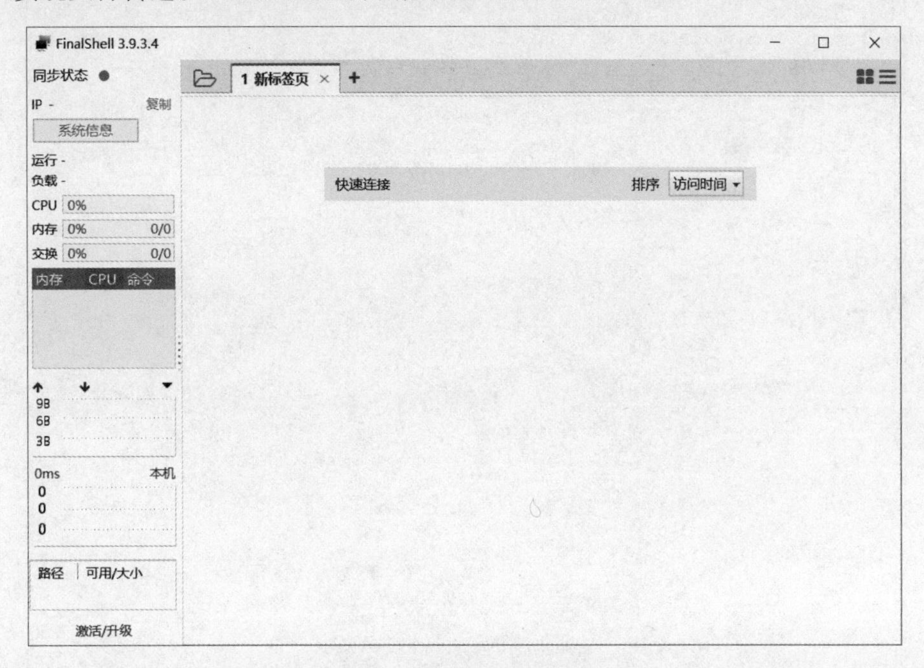

图2-31 FinalShell主界面

点击图2-31中的文件夹图标，再点击添加连接图标，如图2-32所示。

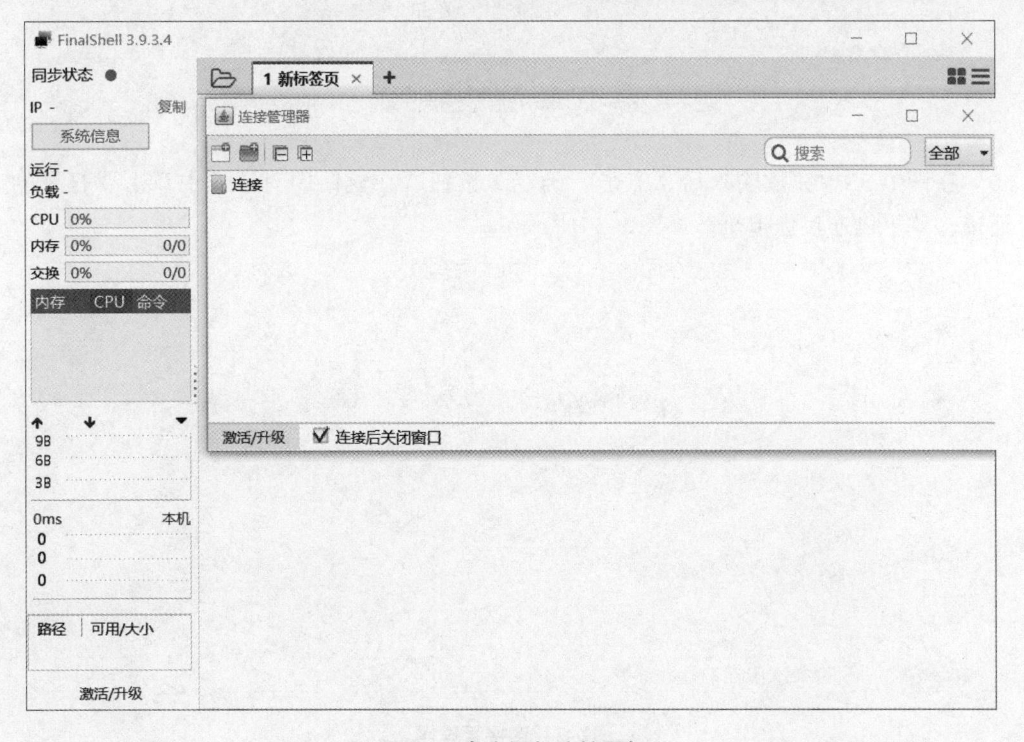

图2-32 点击添加连接图标

在弹出的菜单中选择SSH连接，输入CentOS虚拟机信息即可远程登录虚拟机，如图2-33所示。

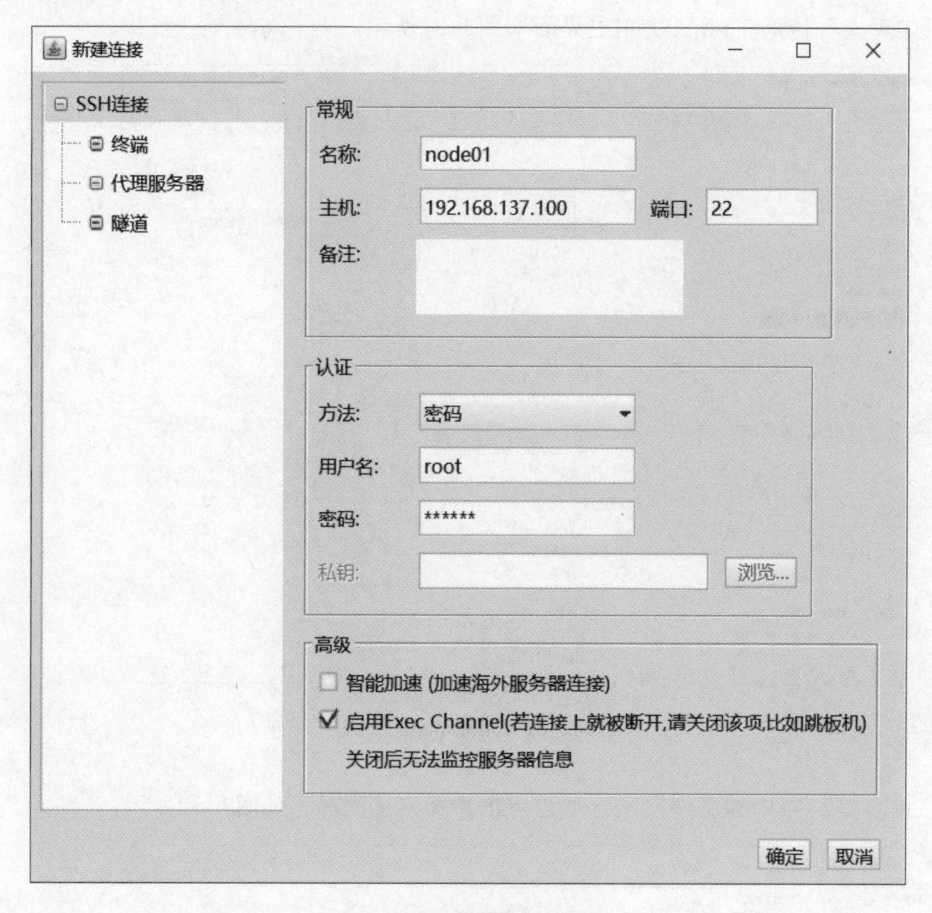

图2-33 填写新建连接信息

在图2-33中完成输入后，点击“确定”按钮，在返回的对话框中双击要打开的连接，就可以连接虚拟机，如图2-34所示。

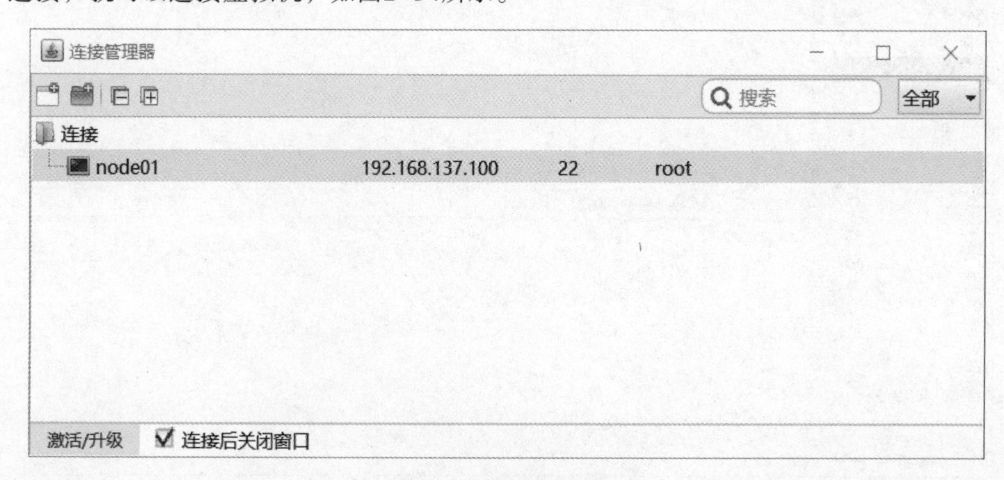

图2-34 连接管理器

连接后的界面如图2-35所示。

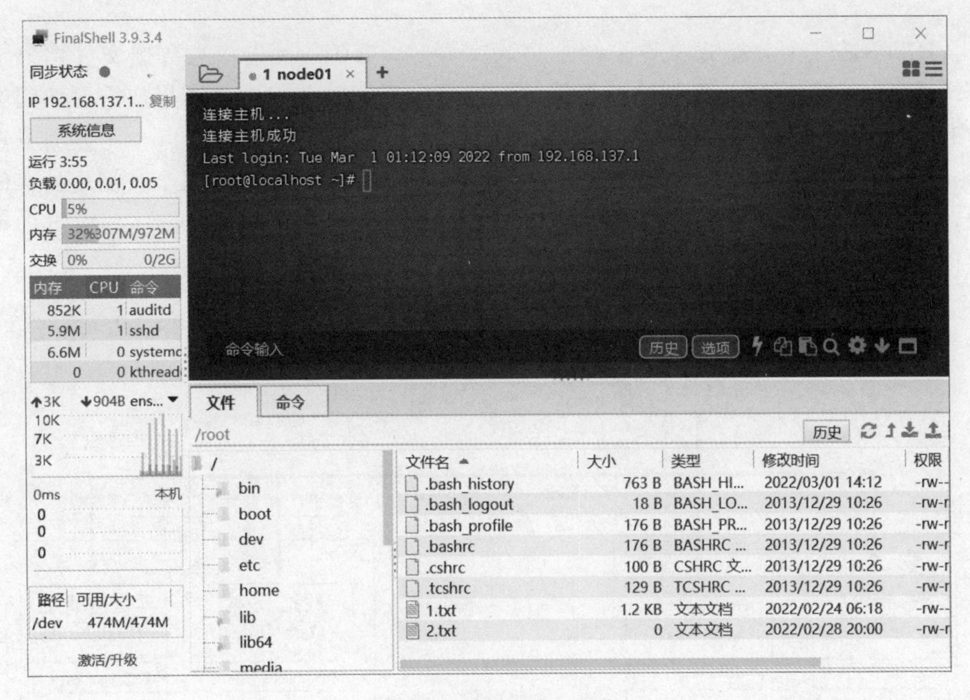

图2-35 连接后的界面

由图2-35可知，图的左侧显示的是系统信息，图的右上部分显示的是命令行界面，图的右下部分显示的是文件管理器，可以用拖曳操作实现文件的上传、下载。

2.2 Hadoop核心组件

Hadoop 2.0框架最核心的设计就是HDFS、MapReduce和YARN，为海量的数据提供了存储和计算，如图2-36所示。

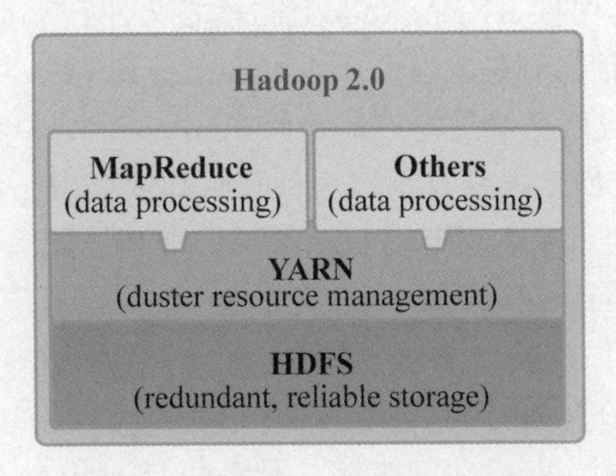

图2-36 Hadoop 2.0框架示意图

2.2.1 HDFS

HDFS是一个分布式文件系统。存入文件时，HDFS首先将大数据文件切分成若干个更小的数据块，再把这些数据块分别写入不同的节点中。当用户需要访问文件时，为了保证能够读取每个数据块，HDFS专门使用集群中的节点（元数据节点NameNode）来保存文件的属性信息，包括文件名、所在目录以及每个数据块的存储位置等，这样，客户端通过NameNode可获得数据块的位置，直接访问DataNode即可获得数据，HDFS架构如图2-37所示。

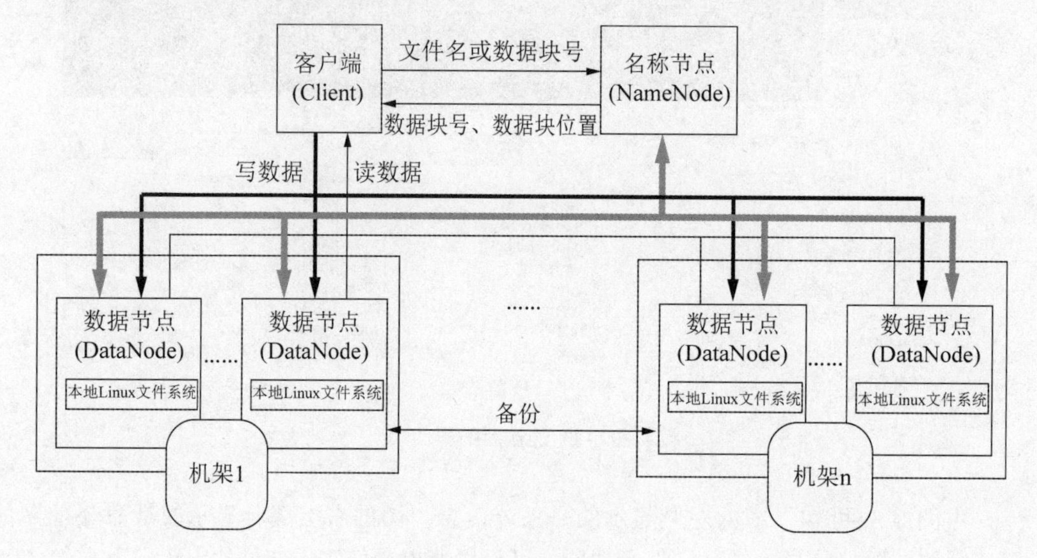

图2-37 HDFS架构

1. Client（客户端）

客户端主要用于上传/下载集群存储的文件，以及操作元数据，常用的有HDFS Web UI、HDFS Shell命令等。

2. NameNode

NameNode在HDFS项目中是管理节点。其本质是想通过fsimage文件和edits文件来实现管理功能。

fsimage文件与edits文件是NameNode上的核心文件。NameNode中仅存储目录树信息，而关于块的位置信息则是从各个DataNode上传到NameNode上。

NameNode的目录树信息就是物理地存储在fsimage文件中，当NameNode启动时会先读取fsimage这个文件，再将目录树信息装载到内存中。而edits文件存储的是日志信息，在NameNode启动后，所有对目录结构的增加、删除、修改等操作都会记录到edits文件中，并不会同步记录到fsimage文件中。

而当NameNode关闭的时候，也不会将fsimage文件与edits文件进行合并，这个合并的过程实际上是发生在NameNode启动的过程中。

也就是说，当NameNode启动的时候，首先装载fsimage文件，然后应用edits文件，其次将最新的目录树信息更新到新的fsimage文件中，最后启用新的edits文件。

3. DataNode

DataNode是集群中实际存储元数据的节点。

4. Block

DataNode在存储数据的时候是以Block（数据块）为单位读/写数据的。Block是HDFS读/写数据的基本单位。假设文件大小为50 GB，从字节位置0开始，每128 MB划分为一个Block，依此类推，可以划分出很多个Block。每个Block就是128 MB大小。

Block本质上是一个逻辑概念，意味着Block里不会真的存储数据，只是划分文件。为防止Block损坏、丢失等影响文件的完整性，HDFS采用的策略是副本机制，让Block在其他节点保存副本。该机制的优点是安全，缺点是占空间，如图2-38所示。

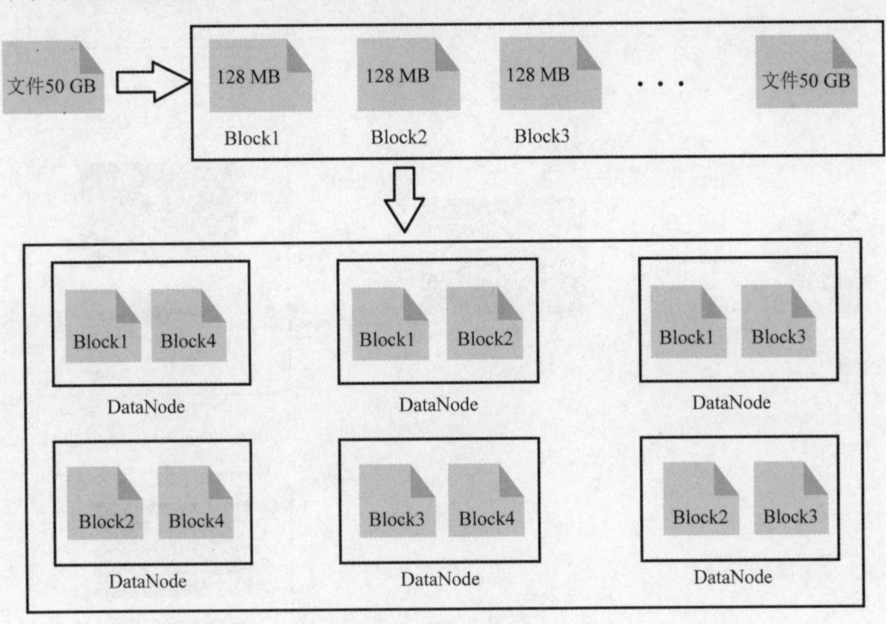

图2-38 Block示意图

5. Secondary NameNode

从NameNode上下载元数据信息（fsimage,edits），然后将二者合并，生成新的fsimage，保存在本地，并将其推送到NameNode，同时重置NameNode的edits。

2.2.2 MapReduce

MapReduce是Hadoop的一个计算核心，由Mapper与Reducer构成。MapReduce的思想就是“分而治之”或者“化繁为简”。

Mapper负责“分”，即把复杂的任务分解为若干个“简单的任务”来处理。

Reducer主要负责将Mapper阶段的结果进行汇总。

2.2.3 YARN

YARN是一种新的Hadoop资源管理器，是一个通用的资源管理系统，可为上层应用提供统一的资源管理和调度，它的引入为集群在利用率、资源统一管理和数据共享等方面带来了巨大好处。YARN由一个全局的资源管理器ResourceManager和每个应用程序特有的ApplicationMaster组成，架构如图2-39所示。

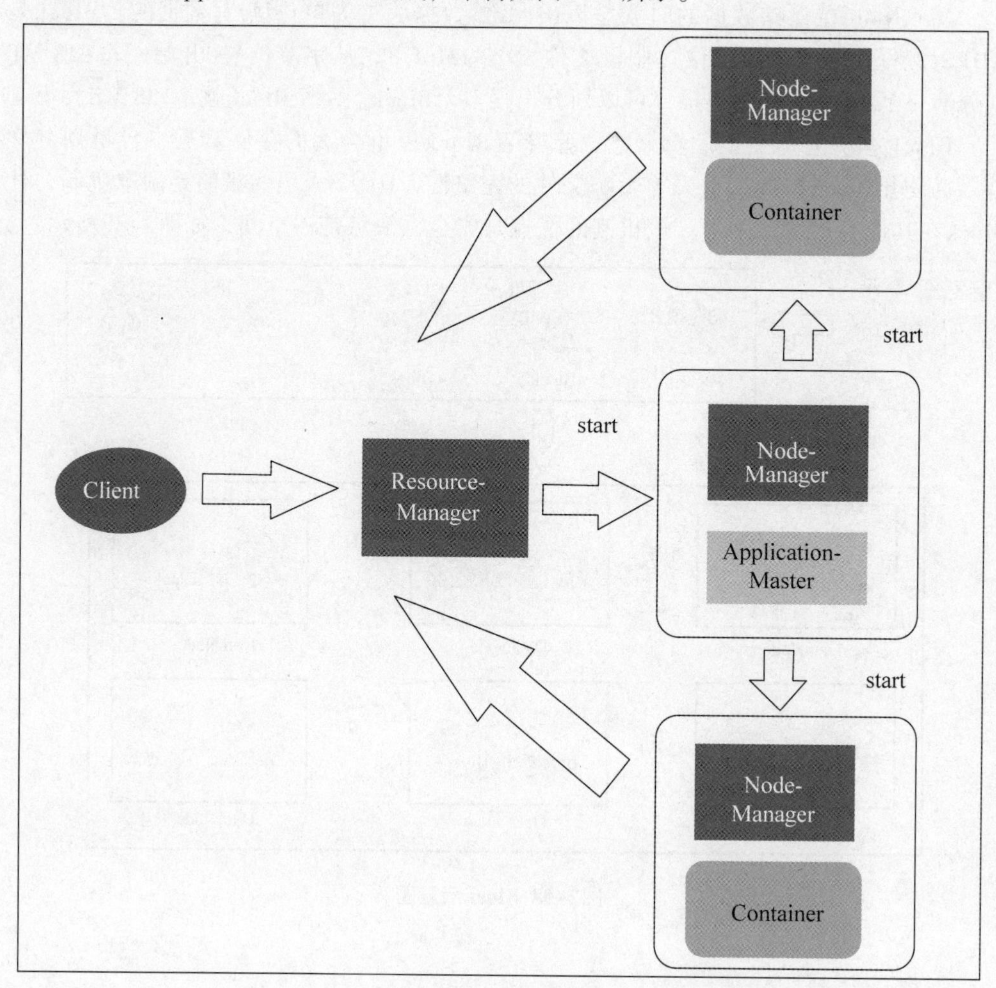

图2-39 YARN架构图

YARN总体上仍然是master/slave结构，在整个资源管理框架中，ResourceManager为master，NodeManager为slave。ResourceManager负责对各个NodeManager上的资源进行统一管理和调度。当用户提交一个计算请求时，需要提供一个用以跟踪和管理这个程序的ApplicationMaster，它负责向ResourceManager申请资源，并要求NodeManager执行可以占用一定资源的任务。由于不同的ApplicationMaster被分布到不同的节点上，因此它们之间不会相互影响。

2.3 Hadoop的搭建

Hadoop有三种运行模式，分别是单机模式（Standalone Mode）、伪分布式模式（Pseudo-Distributed Mode）和全分布式集群模式（Full-Distributed Mode）。

2.3.1 配置准备

Hadoop由Java语言开发，Hadoop集群同样依赖Java环境。在搭建Hadoop集群时，需要安装JDK。

（1）将JDK安装包jdk-8u131-linux-x64.tar.gz传输到CentOS 7服务器的/opt目录下。使用FinalShell拖曳即可。

（2）解压安装包到/usr/local，代码如下：

```
[root@localhost ~]#tar -zxvf /opt/jdk-8u131-linux-x64.tar.gz -C/usr/local
```

（3）配置环境变量。

解压JDK后，需要将JDK加入环境变量中。使用vi/etc/profile命令打开profile文件，在文末添加如下内容：

```
export JAVA_HOME=/usr/local/jdk1.8.0_131
export PATH=$PATH:$JAVA_HOME/bin
```

添加内容后保存并退出，再输入source/etc/profile使配置生效。输入java -version指令测试Java是否完成配置，Java配置成功后如图2-40所示。

```
[root@localhost /]# java -version
java version "1.8.0_131"
Java(TM) SE Runtime Environment (build 1.8.0_131-b11)
Java HotSpot(TM) 64-Bit Server VM (build 25.131-b11, mixed mode)
```

图2-40 Java配置成功

2.3.2 关闭防火墙

配置Hadoop集群需要关闭系统的防火墙，CentOS 7系统默认会关闭iptables，我们关闭firewalld防火墙和selinux防火墙即可，代码如下所示：

```
[root@localhost ~]# systemctl stop firewalld
[root@localhost ~]# systemctl disable firewalld
[root@localhost ~]# systemctl status firewalld
firewalld.service - firewalld - dynamic firewall daemon
   Loaded:loaded(/usr/lib/systemd/system/firewalld.service;disabled)
   Active:inactive(dead)
Oct 12 22:54:57 hadoop systemd[1]:Stopped firewalld -
   dynamic firewall daemon.
```

在关闭firewalld防火墙之前，我们需要先执行systemctl stop firewalld来停止防火墙，避免其处于运行状态，导致关闭失败；然后执行systemctl disable firewalld命令让其彻底不可用；最后执行systemctl status firewalld命令查看防火墙是否关闭成功。从上述代码中可以看到，最后的防火墙状态是inactive(dead)，说明操作成功。

2.3.3 本地模式的环境搭建

微课：
v2-6 配置单机Hadoop

Hadoop本地模式（Local Mode）的安装，即是Hadoop单机运行，存储和运算都在当前的Linux主机运行，常用于开发工程中的业务逻辑调试。

1. 解压Hadoop

将Hadoop安装包hadoop-2.7.7.tar.gz传输到CentOS 7服务器的/opt目录下，再使用tar命令解压Hadoop安装包到/usr/local文件夹下：

```
[root@localhost opt]# tar -zxf /opt/hadoop-2.7.7.tar.gz -C /usr/local/
[root@localhost opt]# ls /usr/local/
bin etc games hadoop-2.7.7 include jdk1.8.0_131 lib lib64 libexec sbin
share sr
```

在配置Hadoop之前，还需要了解目录结构是怎么样的，使用cd命令进入hadoop-2.7.7目录，如下所示：

```
[root@localhost hadoop-2.7.7]# ll
totle 112
drwxr-xr-x. 2 1000 ftp   194 7月  18 2018 bin
drwxr-xr-x. 3 1000 ftp    20 7月  18 2018 etc
drwxr-xr-x. 2 1000 ftp   106 7月  18 2018 include
drwxr-xr-x. 3 1000 ftp    20 7月  18 2018 lib
drwxr-xr-x. 2 1000 ftp   239 7月  18 2018 libexec
-rw-r--r--. 1 1000 ftp 86424 7月  18 2018 LICENSE.txt
-rw-r--r--. 1 1000 ftp 14978 7月  18 2018 NOTICE.txt
-rw-r--r--. 1 1000 ftp  1366 7月  18 2018 README.txt
drwxr-xr-x. 2 1000 ftp  4096 7月  18 2018 sbin
drwxr-xr-x. 4 1000 ftp    31 7月  18 2018 share
```

Hadoop目录结构说明如表2-3所示。

表2-3 Hadoop目录结构说明

目录名称	描 述
bin	可执行文件，用于存放常用指令
etc	Hadoop配置文件所在位置，具体为etc/hadoop
lib	Hadoop运行依赖的第三方包
Share	Hadoop文档和示例
sbin	一些可执行脚本，比如start-dfs.sh、stop-dfs.sh等启动和停止命令
logs	Hadoop运行时产生的日志文件
tmp	格式化HDFS文件系统时产生的目录
include	namespace文件、工具文件
libexec	一些可执行的脚本配置文件

2.配置环境变量

将Hadoop安装包解压后，需要对Hadoop进行配置。

将Hadoop的安装目录、指令文件夹加入环境变量中，可以方便后续的操作。打开环境变量配置文件/etc/profile进行编辑，在配置好jdk的基础上添加如下代码：

```
[root@localhost local]# vi/etc/profile
…
export JAVA_HOME=/usr/local/jdk1.8.0_131
export HADOOP_HOME=/usr/local/hadoop-2.7.7
export PATH=$PATH:$JAVA_HOME/bin:$HADOOP_HOME/bin:$HADOOP_HOME/sbin
```

执行source/etc/profile命令来更新环境变量配置文件。可以输入hadoop指令测试配置情况，代码如下所示：

```
[root@localhost hadoop-2.7.7]# hadoop version
Hadoop 2.7.7
Subversion Unknown -r claad84bd27cd79c3d1a7dd58202a8c3ee1ed3ac
Compiled by stevel on 2018-07-18T22:47Z
Compiled with protoc 2.5.0
From source with checksum 792e15d20b12c74bd6f19a1fb886490
This command was run using/usr/local/hadoop-2.7.7/share/hadoop/common/
    hadoop-common-2.7.7.jar
```

3.配置Hadoop运行环境

由于Hadoop需要依赖JDK环境，需要在hadoop-env.sh文件中配置JDK安装路径，因此，需要将文件中的export JAVA_HOME= $ {JAVA_HOME}修改成实际的JDK安装路径，代码如下所示：

```
[root@localhost ~]# vi $HADOOP_HOME/etc/hadoop/hadoop-env.sh
#export JAVA_HOME=${JAVA_HOME}
export JAVA_HOME=/usr/local/jdk1.8.0_131
```

此时，单机版的Hadoop配置完毕，可以通过Hadoop的示例程序pi（采用Quasi-Monte Carlo算法来估算pi的值）来测试单机版的Hadoop是否配置成功，代码如下：

```
hadoop jar $HADOOP_HOME/share/hadoop/mapreduce/hadoop-mapreduce-exam
ples-2.7.7.jar pi 10 10
```

可以看到执行后的日志如下所示：

```
Number of Maps  = 10
Samples per Map = 10
Wrote input for Map #0
Wrote input for Map #1
Wrote input for Map #2
Wrote input for Map #3
Wrote input for Map #4
Wrote input for Map #5
```

```
Wrote input for Map #6
Wrote input for Map #7
Wrote input for Map #8
Wrote input for Map #9
Starting Job
22/03/15 08:37:12 INFO Configuration.deprecation:session.id is
    deprecat ed.Instead,use dfs.metrics.session-id
22/03/15 08:37:12 INFO jvm.JvmMetrics:Initializing JVM Metrics with
    processName = JobTracker,sessionId=
    22/03/15 08:37:12 INFO input.FileInputFormat:Total input paths to
    pro cess:10
22/03/15 08:37:12 INFO mapreduce.JobSubmitter:number of splits:10
22/03/15 08:37:13 INFO mapreduce.JobSubmitter:Submitting tokens for
    job:job_local1285704942_0001
22/03/15 08:37:13 INFO mapreduce.Job:The url to track the job:http://
    localhost:8080/
22/03/15 08:37:13 INFO mapreduce.Job:Running job:
    job_local1285704942_0001
22/03/15 08:37:13 INFO mapred.LocalJobRunner:OutputCommitter set in
    config null
22/03/15 08:37:13 INFO output.FileOutputCommitter:File Output Committer
    Algorithm version is 1
22/03/15 08:37:13 INFO mapred.LocalJobRunner:OutputCommitter is
    org.apache.hadoop.mapreduce.lib.output.FileOutputCommitter
22/03/15 08:37:13 INFO mapred.LocalJobRunner:Waiting for map tasks
22/03/15 08:37:13 INFO mapred.LocalJobRunner:Starting task:
    attempt_local1285704942_0001_m_000000_0
22/03/15 08:37:13 INFO output.FileOutputCommitter:File Output Committer
    Algorithm version is 1
22/03/15 08:37:13 INFO mapred.Task:Using
    ResourceCalculatorProcessTree:[]
22/03/15 08:37:13 INFO mapred.MapTask:Processing split:file:/usr/local/
    hadoop-2.7.7/QuasiMonteCarlo_1647347830846_1598835670/in/part0:0+118
22/03/15 08:37:13 INFO mapred.MapTask:(EQUATOR)0 kvi 26214396(104857584)
22/03/15 08:37:14 INFO mapred.MapTask:mapreduce.task.io.sort.mb:100
22/03/15 08:37:14 INFO mapred.MapTask:soft limit at 83886080
22/03/15 08:37:14 INFO mapred.MapTask:bufstart = 0;bufvoid = 104857600
22/03/15 08:37:14 INFO mapred.MapTask:kvstart = 26214396;length =
    6553600
22/03/15 08:37:14 INFO mapred.MapTask:Map output collector class =
    org.apache.hadoop.mapred.MapTask$MapOutputBuffer
22/03/15 08:37:14 INFO mapred.LocalJobRunner:
22/03/15 08:37:14 INFO mapred.MapTask:Starting flush of map output
22/03/15 08:37:14 INFO mapred.MapTask:Spilling map output
22/03/15 08:37:14 INFO mapred.MapTask:bufstart = 0;bufend = 18;bufvoid
    = 104857600
22/03/15 08:37:14 INFO mapred.MapTask:kvstart =
    26214396(104857584);kvend = 26214392(104857568);length = 5/6553600
22/03/15 08:37:14 INFO mapred.MapTask:Finished spill 0
22/03/15 08:37:14 INFO mapred.Task:Task:
```

```
    attempt_local1285704942_0001_m_000000_0 is done.And is in the process of
    committing
22/03/15 08:37:14 INFO mapred.LocalJobRunner:map
22/03/15 08:37:14 INFO mapred.Task:Task
    'attempt_local1285704942_0001_m_000000_0' done.
22/03/15 08:37:14 INFO mapred.Task:Final Counters for attempt_local1285
    704942_0001_m_000000_0:Counters:17
22/03/15 08:37:15 INFO mapred.Task:Final Counters for attempt_local1285
    704942_0001_r_000000_0:Counters:24
22/03/15 08:37:15 INFO mapred.LocalJobRunner:Finishing task:
    attempt_local1285704942_0001_r_000000_0
22/03/15 08:37:15 INFO mapred.LocalJobRunner:reduce task executor
    complete.
22/03/15 08:37:15 INFO mapreduce.Job:map 100% reduce 100%
22/03/15 08:37:15 INFO mapreduce.Job:Job job_local1285704942_0001
    completed successfully
22/03/15 08:37:15 INFO mapreduce.Job:Counters:30
    File System Counters
        FILE:Number of bytes read=3345940
        FILE:Number of bytes written=6525473
        FILE:Number of read operations=0
        FILE:Number of large read operations=0
        FILE:Number of write operations=0
    Map-Reduce Framework
        Map input records=10
        Map output records=20
        Map output bytes=180
        Map output materialized bytes=280
        Input split bytes=1430
        Combine input records=0
        Combine output records=0
        Reduce input groups=2
        Reduce shuffle bytes=280
        Reduce input records=20
        Reduce output records=0
        Spilled Records=40
        Shuffled Maps =10
        Failed Shuffles=0
        Merged Map outputs=10
        GC time elapsed (ms)=292
        Total committed heap usage (bytes)=1677926400
    Shuffle Errors
        BAD_ID=0
        CONNECTION=0
        IO_ERROR=0
        WRONG_LENGTH=0
        WRONG_MAP=0
        WRONG_REDUCE=0
    File Input Format Counters
        Bytes Read=1300
    File Output Format Counters
```

```
        Bytes Written=109
Job Finished in 3.498 seconds
Estimated value of Pi is 3.20000000000000000000
```

从运行日志中不难发现，程序启动了一个Job(Running job:job_local1285704942_0001)，从任务编号可以得出，当前Hadoop是以local模式运行的。由于计算时pi的参数是10 10，代表运行10个map任务，每个任务投掷10次参数，因此计算精度不高。读者可以自行增大数字，以提升计算结果的精度。

在完成单机版的Hadoop配置后，需要使用VMware的快照功能来记录当前状态。当后续操作出现意外时，可以即时恢复到单机版的Hadoop配置成功的状态。

2.3.4 伪分布式模式

微课：v2-7 配置伪分布式_Trim

单机模式在一台单机上运行，没有分布式文件系统，而是直接读/写本地操作系统的文件系统。默认情况下，Hadoop被配置成以非分布式模式运行的一个独立Java进程。而全分布式Hadoop的搭建步骤较为烦琐，性能消耗大，编写程序时，更多的是搭建伪分布式Hadoop集群。

1. 免密登录

全分布式Hadoop是由多台CentOS服务器形成一个集群，将管理节点与数据节点分别配置在各台服务器中。而伪分布式集群则是将管理节点、数据节点等所有进程都安装到一台服务器中，以一种自己管理自己、自己调度自己的模式运行，因此需要将服务器设置为自己的免密登录。在全分布式中，则是服务器之间的免密登录。生成的密钥代码如下所示：

```
[root@localhost ~]# ssh-keygen -t dsa -P '' -f ~/.ssh/id_dsa
Generating public/private dsa key pair.
Created directory '/root/.ssh'.
Your identification has been saved in /root/.ssh/id_dsa.
Your public key has been saved in /root/.ssh/id_dsa.pub.
The key fingerprint is:
SHA256:kPUNcU96YKGYaTkc15GKw3d9hPwR2PrGGZjcKmFah1o root@192.168.137.100
The key's randomart image is:
+---[DSA 1024]----+
|        o +o*=+o.|
|       + O *o*oo.|
|      o.O.o.=.Bo.|
|       o+.oE.B.oo|
|        So*.o +.o|
|         o . . = |
|            . .  |
|                 |
|                 |
+----[SHA256]-----+
```

生成密钥后，需要将生成的密钥复制到.ssh/authorized_keys中，代码如下所示：

```
[root@localhost ~]# cat ~/.ssh/id_dsa.pub >> ~/.ssh/authorized_keys
```

使用ssh localhost命令可以测试对本机的免密登录，使用logout可以退出远程登录模式。

2. core-site.xml文件配置

core-site.xml文件为Hadoop的核心配置文件，进入安装目录下的etc/hadoop，配置后的文件代码如下所示：

```
[root@localhost ~] #cd $HADOOP_HOME/etc/hadoop/
[root@localhost hadoop] #vi core-site.xml
...
<configuration>
    <property>
            <name>fs.defaultFS</name>
            <value>hdfs://localhost:9000</value>
        </property>
        <property>
            <name>hadoop.tmp.dir</name>
            <value>/opt/hadoop/hadoop-2.7.7/tmp</value>
    </property>
</configuration>
```

在上面的代码中，主要配置了两个属性：第一个属性用于指定HDFS的NameNode的通信地址，这里我们将其指定为Hadoop；第二个属性用于指定Hadoop运行时产生的文件存放目录，这个目录我们不需要去创建，因为在格式化Hadoop的时候会自动创建。

3. hdfs-site.xml文件配置

hdfs-site.xml文件为HDFS的核心配置文件，配置后的文件代码如下所示：

```
[root@localhost ~] #cd $HADOOP_HOME/etc/hadoop/
[root@hadoop hadoop]#vi hdfs-site.xml
...
<configuration>
    <property>
        <name>dfs.replication</name>
        <value>1</value>
    </property>
    <property>
        <name>dfs.namenode.name.dir</name>
        <value>/usr/local/hadoop-2.7.7/tmp/name</value>
    </property>
    <property>
        <name>dfs.datanode.data.dir</name>
        <value>/usr/local/hadoop-2.7.7/tmp/data</value>
    </property>
</configuration>
```

Hadoop集群的默认副本数是3，但是，现在我们只是在单节点上进行伪分布式安装，不需要保存3个副本，我们将该属性的值修改为1即可。

4. mapred-site.xml文件配置

mapred-site.xml文件是不存在的，但是有一个模板文件mapred-site.xml.template，我们将模板文件改名为mapred-site.xml，然后进行编辑。mapred-site.xml文件为MapReduce的核心配置文件，配置后的文件代码如下所示：

```
[root@localhost ~] #cd $HADOOP_HOME/etc/hadoop/
[root@hadoop hadoop] #cp mapred-site.xml.template mapred-site.xml
[root@hadoop hadoop] #vi mapred-site.xml
...
<configuration>
    <property>
        <name>mapreduce.framework.name</name>
        <value>yarn</value>
    </property>
</configuration>
```

之所以配置上面的属性，是因为在Hadoop 2.0之后，MapReduce是运行在YARN框架上的，需要进行特别声明。

5. yarn-site.xml文件配置

yarn-site.xml文件为YARN的框架配置文件，主要用于指定ResourceManager的节点名称及NodeManager属性，配置后的文件代码如下所示：

```
[root@localhost ~] #cd $HADOOP_HOME/etc/hadoop/
[root@hadoop hadoop] #vi yarn-site.xml
...
<configuration>
    <property>
        <name>yarn.resourcemanager.hostname</name>
        <value>localhost</value>
    </property>
    <property>
        <name>yarn.nodemanager.aux-services</name>
        <value>mapreduce_shuffle</value>
    </property>
</configuration>
```

在上面的代码中，我们配置了两个属性：第一个属性用于指定ResourceManager的地址，因为是单节点部署，所以指定为localhost即可；第二个属性用于指定Reducer获取数据的方式。

6. 格式化文件系统

Hadoop集群是主从结构，分布式文件系统采用的是NameNode管理DataNode。在启动之前需要对NameNode进行格式化，代码如下所示：

```
[root@localhost ~] #hdfs NameNode -format
```

格式化成功后，在Hadoop中会自动生成tmp目录，用于存放Hadoop运行中的临时文件。

注意，在两次格式化NameNode后，就会产生新的集群clusterID，导致NameNode和DataNode的clusterID不一致，集群找不到以前的数据。如果要格式化NameNode，在格式化之前，一定要先删除data数据和log日志（如果是在搭建好集群后格式化，那就要把所有的日志都清空，包括Zookeeper，但最好不要这样做），然后再格式化NameNode。

7. HDFS启动

HDFS启动指令如下所示：

```
[root@localhost ~]# start-dfs.sh
```

在启动HDFS时，由于sbin目录已经加入环境变量中，可以直接在任何目录下输入start-dfs.sh完成启动。在第一次启动时，由于没有设置对ip0.0.0.0的免密登录，所以会提示是否继续连接，输入yes，完成HDFS的启动，显示代码如下所示：

```
Starting namenodes on [localhost]
localhost: starting namenode, logging to /usr/local/hadoop-2.7.7/logs/
hadoop-root-namenode-192.168.137.100.out
localhost: starting datanode, logging to /usr/local/hadoop-2.7.7/logs/
hadoop-root-datanode-192.168.137.100.out
Starting secondary namenodes [0.0.0.0]
The authenticity of host '0.0.0.0 (0.0.0.0)' can't be established.
ECDSA key fingerprint is SHA256:
    9k0J6CA6x0zoZvuJRb2qwHfYKpodCV564ZIkGyg 4VCU.
ECDSA key fingerprint is MD5:22:cc:d0:5c:46:e2:77:5b:c4:f6:
    b4:13:d2:2a:50:82.
Are you sure you want to continue connecting (yes/no)
```

输入jps命令可以查看HDFS相关进程是否启动成功，代码如下所示：

```
[root@localhost ~]# jps
17107 NameNode
17284 DataNode
20664 SecondaryNameNode
21865 jps
```

从jps的执行结果来看，NameNode、DataNode、SecondaryNameNode成功启动。当启动失败时，可以针对未启动进程查看相关文件是否配置成功。删除tmp文件夹中的所有内容后，重新格式化NameNode，再重新启动HDFS。

输入如下命令可以查看HDFS集群状态：

```
[root@localhost ~] #hdfs dfsadmin -report
Configured Capacity:18238930944 (16.99 GB)
Present Capacity:15787200512 (14.70 GB)
DFS Remaining:15787192320 (14.70 GB)
```

```
DFS Used:8192 (8 KB)
DFS Used%:0.00%
Under replicated blocks:0
Blocks with corrupt replicas:0
Missing blocks:0
Missing blocks (with replication factor 1):0
-------------------------------------------------
Live datanodes (1):
Name:127.0.0.1:50010 (localhost)
Hostname:192.168.137.100
Decommission Status:Normal
Configured Capacity:18238930944 (16.99 GB)
DFS Used:8192 (8 KB)
Non DFS Used:2451730432 (2.28 GB)
DFS Remaining: 15787192320 (14.70 GB)
DFS Used%:0.00%
DFS Remaining%:86.56%
Configured Cache Capacity:0 (0 B)
Cache Used:0 (0 B)
Cache Remaining:0 (0 B)
Cache Used%:100.00%
Cache Remaining%:0.00%
Xceivers:1
Last contact:Wed Mar 16 07:55:04 EDT 2022
```

或者通过Hadoop的可视化管理界面查看HDFS的运行情况。在Windows主机中打开浏览器，在地址栏输入http://192.168.137.100:50070访问HDFS可视化管理界面，如图2-41所示。

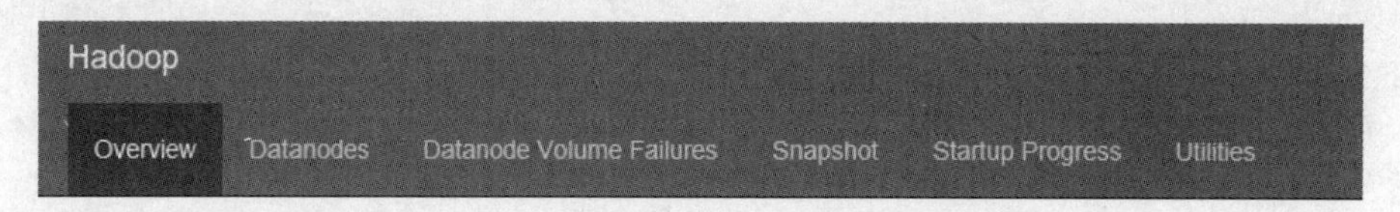

Overview 'localhost:9000' (active)

Started:	Wed Mar 16 07:37:08 EDT 2022
Version:	2.7.7, rc1aad84bd27cd79c3d1a7dd58202a8c3ee1ed3ac
Compiled:	2018-07-18T22:47Z by stevel from branch-2.7.7
Cluster ID:	CID-9a90aefe-6b3e-4003-9595-ed0d10c919dd
Block Pool ID:	BP-1675166625-192.168.137.140-1647430324188

图2-41 HDFS可视化管理界面

8. YARN启动

YARN启动命令如下所示：

```
[root@localhost ~]# start-yarn.sh
```

输入jps命令可以查看YARN的相关进程是否启动成功，代码如下所示：

```
[root@localhost ~] #jps
17107 NameNode
17284 DataNode
20664 SecondaryNameNode
31416 ResourceManager
31693 jps
31550 NodeManager
```

通过jps可以看到，在启动YARN的过程中，分别启动了ResourceManager、NodeManager。

在Windows主机中打开浏览器，在地址栏输入http://192.168.137.100:8088访问YARN可视化管理界面，如图2-42所示。

All Applications

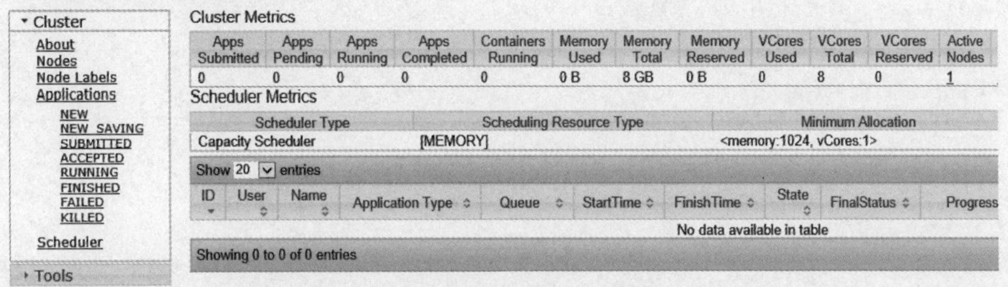

图2-42 YARN可视化管理界面

通过YARN可视化管理界面，可以知道系统在执行一项任务的时候启动了多少个Job，监听每个Job的运行资源情况，还可以查看Job的历史，以及哪些Job运行成功、哪些Job运行失败。

9. 关闭HDFS

先关闭HDFS，代码如下所示：

```
[root@localhost ~]# stop-dfs.sh
Stopping namenodes on [localhost]
localhost:stopping namenode
localhost:stopping datanode
Stopping secondary namenodes [0.0.0.0]
0.0.0.0:stopping secondarynamenode
```

10. 关闭YARN

再关闭YARN，代码如下所示：

```
[root@localhost ~] #stop-yarn.sh
stopping yarn daemons
stopping resourcemanager
localhost:stopping nodemanager
no proxyserver to stop
```

2.3.5 全分布式模式

微课：
v2-8 全分布式Hadoop

全分布式Hadoop与多台CentOS服务器形成一个集群，将管理节点与数据节点分别配置在各台服务器中。本节将在伪分布式Hadoop的基础上配置相关文件，再克隆虚拟机，完成集群的搭建。集群规划如表2-4所示。

表2-4 集群规划

主机名	IP	HDFS进程	YARN进程
master	192.168.137.100	NameNode、SecondaryNameNode	ResourceManager
slave1	192.168.137.101	DataNode	NodeManager
slave2	192.168.137.102	DataNode	NodeManager

1. 配置core-site.xml文件

修改伪分布式配置中的core-site.xml文件，代码如下所示：

```
[root@localhost ~] #vi $HADOOP_HOME/etc/hadoop/core-site.xml
<configuration>
<property>
    <name>fs.defaultFS</name>
        <value>hdfs://master:9000</value>
    </property>
    <property>
        <name>hadoop.tmp.dir</name>
        <value>/usr/local/hadoop-2.7.7/tmp</value>
    </property>
</configuration>
```

根据集群规划，NameNode节点将部署在master主机上，因此，修改HDFS的NameNode通信地址。

2. 修改hdfs-site.xml文件

调整HDFS的副本为2，代码如下所示：

```
[root@localhost ~] #vi $HADOOP_HOME/etc/hadoop/hdfs-site.xml
<configuration>
    <property>
        <name>dfs.replication</name>
        <value>2</value>
    </property>
    <property>
        <name>dfs.namenode.name.dir</name>
        <value>/usr/local/hadoop-2.7.7/tmp/name</value>
    </property>
    <property>
        <name>dfs.datanode.data.dir</name>
        <value>/usr/local/hadoop-2.7.7/tmp/data</value>
    </property>
</configuration>
```

3. 修改yarn-site.xml文件

配置YARN支持实时计算，修改核心配置文件yarn-site.xml，代码如下所示：

```
[root@localhost ~] #vi $HADOOP_HOME/etc/hadoop/yarn-site.xml
<configuration>
    <property>
        <name>yarn.resourcemanager.hostname</name>
        <value>master</value>
    </property>
    <property>
        <name>yarn.nodemanager.aux-services</name>
        <value>mapreduce_shuffle</value>
    </property>
</configuration>
```

4. 修改slaves

为集群设置从节点，设置DataNode和NodeManager。删除slaves文件原数据，在文末添加从节点host或IP。代码如下所示：

```
[root@localhost ~]# vi $HADOOP_HOME/etc/hadoop/slaves
slave1
slave2
```

5. 修改hosts

对主机名称和IP地址进行映射，对hosts进行配置，代码如下所示：

```
[root@localhost ~]# vi /etc/hosts
192.168.137.100 master
192.168.137.101 slave1
192.168.137.102 slave2
```

6. 清空文件

由于伪分布式虚拟机在操作过程中格式化了NameNode，已经产生了clusterID、data数据和log日志，因此需要删除这些数据，防止对集群造成干扰。需要删除的代码如下所示：

```
[root@localhost ~]# rm -fr $HADOOP_HOME/tmp
[root@localhost ~]# rm -fr $HADOOP_HOME/logs/*
[root@localhost ~]# rm -fr /tmp/*
```

7. 克隆虚拟机

（1）关闭当前node01虚拟机，在左侧列表右击node01节点，在弹出的菜单中选择“快照”→“拍摄快照”，在弹出的“node01 -拍摄快照”窗口的“名称”后的输入框中将快照命名为“配置好全分布式”，如图2-43所示。

（2）在左侧列表右击node01节点，在弹出的菜单中选择“管理”→“克隆”，如图2-44所示。

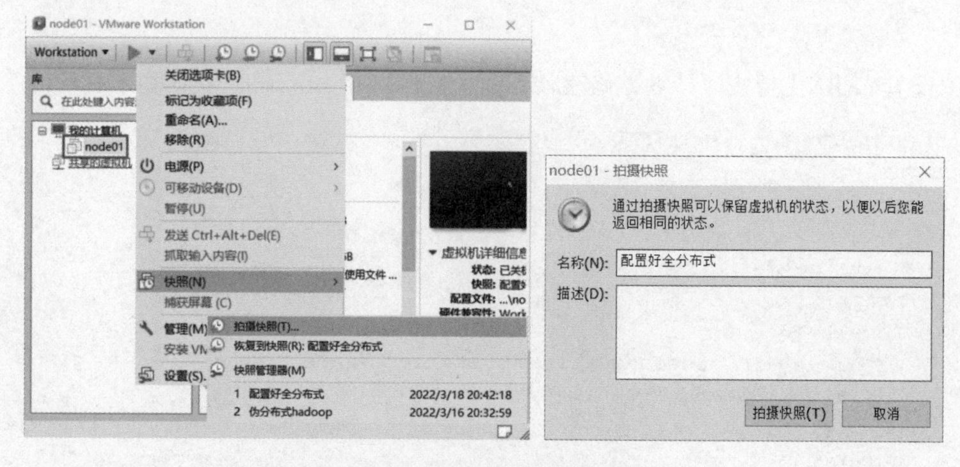

图2-43 “node01 -拍摄快照”窗口

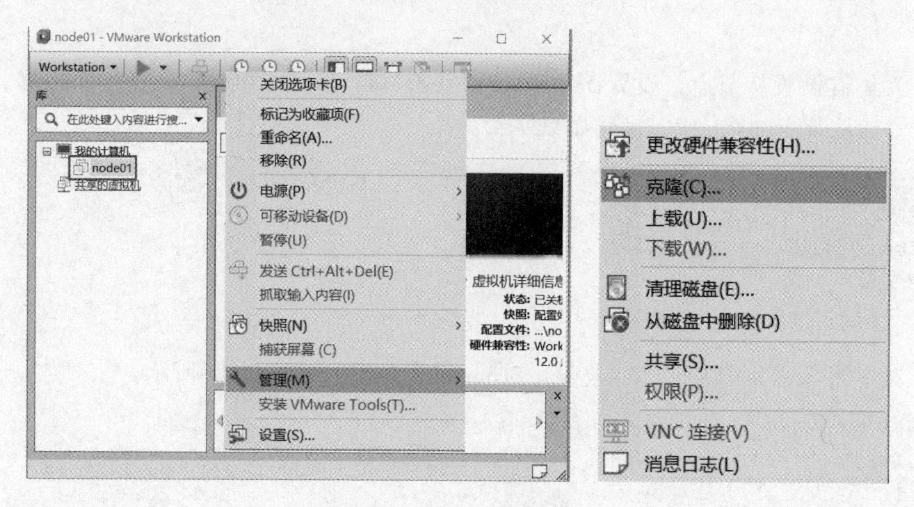

图2-44 选择克隆虚拟机

（3）在弹出的欢迎界面中点击“下一步”按钮，选择“现有快照（仅限关闭的虚拟机）”选项，在下拉菜单中选择“配置好全分布式”，并点击“下一步”按钮确认，如图2-45所示。

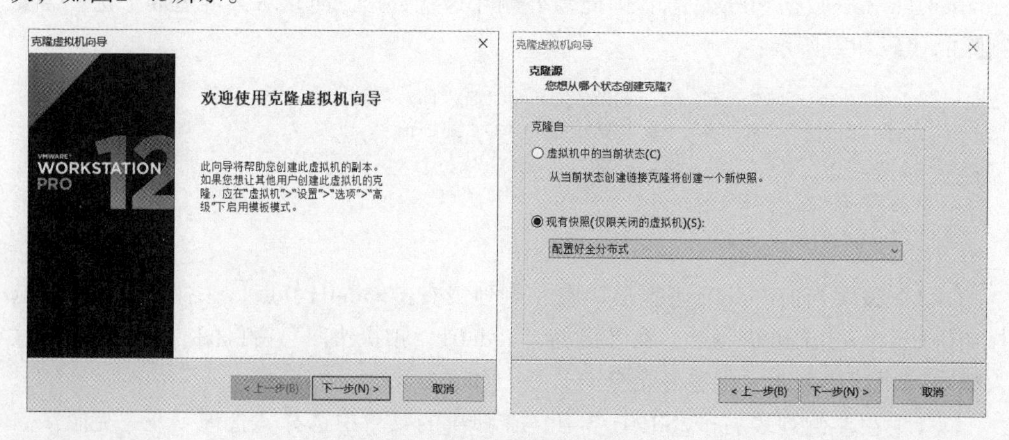

图2-45 选择克隆虚拟机选项

（4）选择“创建完整克隆”选项，并点击“下一步”按钮，设置“虚拟机名称”为“slave1”，“位置”选择自定义的虚拟的存储文件夹，如图2-46所示。点击“完成”按钮，完成slave1的克隆。

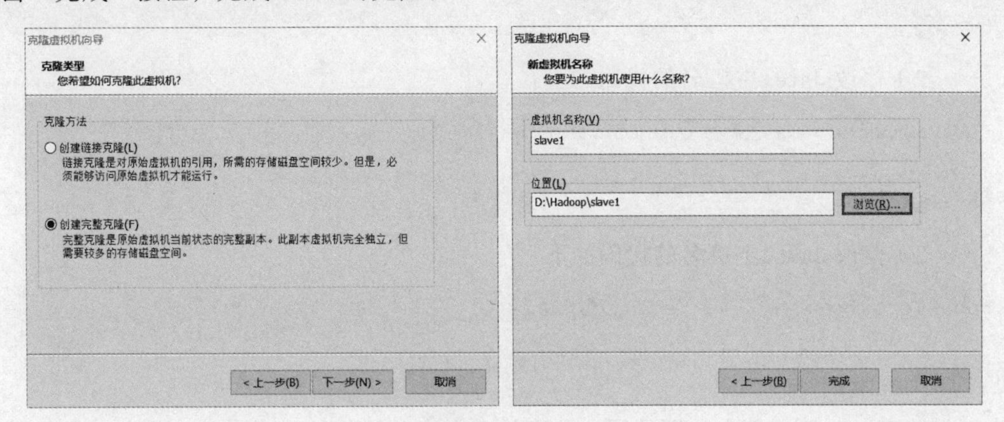

图2-46 虚拟机命名

（5）slave2的克隆操作与slave1的克隆操作相同，同样是从node01主机克隆。完成克隆后的主机列表如图2-47所示。

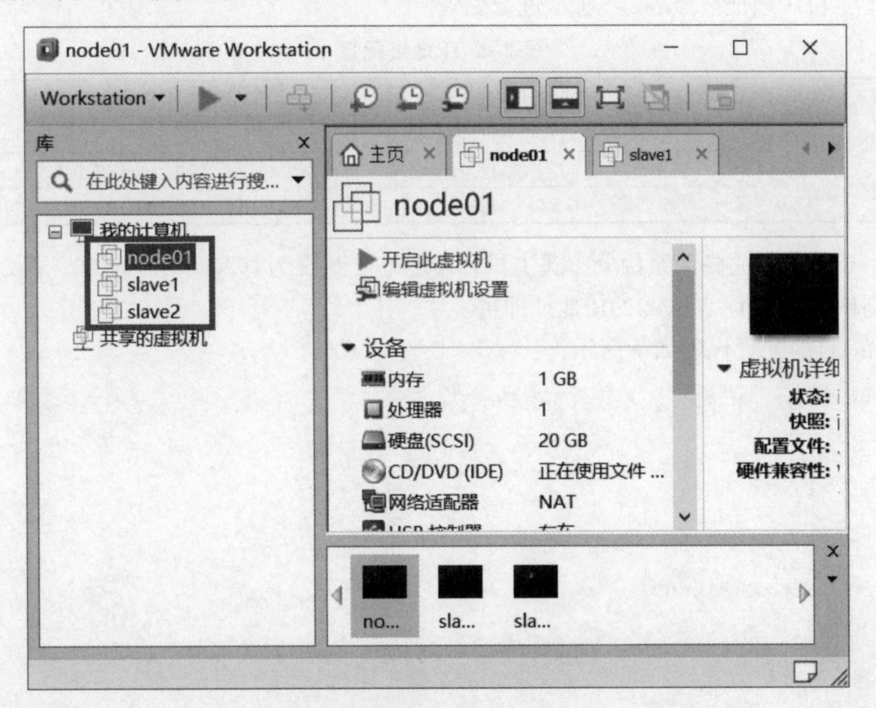

图2-47 完成克隆后的主机列表

8. 修改主机名

分别启动3台虚拟机。按照规划，修改node01主机名为master，修改slave1、slave2主机名为slave1、slave2。

（1）修改master主机名的代码如下：

```
[root@localhost ~] #vi /etc/hostname

master
```

（2）修改slave1主机名的代码如下：

```
[root@localhost ~] #vi /etc/hostname

slave1
```

（3）修改slave2主机名的代码如下：

```
[root@localhost ~] #vi /etc/hostname

slave2
```

删除hostname文件中的源数据，分别将三台虚拟机修改成对应的主机名。修改后需要重启才能生效。

9.配置静态IP地址

将主机IP地址按照表2-5所示进行配置。

表2-5 IP地址配置

虚拟机名	主机名	IP地址
node01	master	192.168.137.100
slave1	slave1	192.168.137.101
slave2	slave2	192.168.137.102

由于master主机是原伪分布式主机，IP地址已经设为192.168.137.100，所以只需要分别修改slave1和slave2的IP地址即可。

slave1的配置代码如下所示：

```
[root@slave1 ~] #vi /etc/sysconfig/network-scripts/ifcfg-ens33

TYPE=Ethernet
PROXY_METHOD=none
BROWSER_ONLY=no
BOOTPROTO=static
DEFROUTE=yes
IPV4_FAILURE_FATAL=no
IPV6INIT=yes
IPV6_AUTOCONF=yes
IPV6_DEFROUTE=yes
IPV6_FAILURE_FATAL=no
IPV6_ADDR_GEN_MODE=stable-privacy
NAME=ens33
UUID=a5211f41-4f33-4f65-8cd6-ebb1e677210d
DEVICE=ens33
ONBOOT=yes
IPADDR=192.168.137.101
GATEWAY=192.168.137.2
NETMASK=255.255.255.0
DNS1=114.114.114.114
```

slave2的配置代码如下所示：

```
[root@slave2 ~] #vi /etc/sysconfig/network-scripts/ifcfg-ens33

TYPE=Ethernet
PROXY_METHOD=none
BROWSER_ONLY=no
BOOTPROTO=static
DEFROUTE=yes
IPV4_FAILURE_FATAL=no
IPV6INIT=yes
IPV6_AUTOCONF=yes
IPV6_DEFROUTE=yes
IPV6_FAILURE_FATAL=no
IPV6_ADDR_GEN_MODE=stable-privacy
NAME=ens33
UUID=a5211f41-4f33-4f65-8cd6-ebb1e677210d
DEVICE=ens33
ONBOOT=yes
IPADDR=192.168.137.102
GATEWAY=192.168.137.2
NETMASK=255.255.255.0
DNS1=114.114.114.114
```

配置完成后，需要输入service network restart命令，重新加载网络配置。可以在三台主机相互ping测试网络是否配置成功。

10. 格式化文件系统

Hadoop集群是主从结构，分布式文件系统采用的是NameNode管理DataNode节点。在启动之前需要对NameNode进行格式化，代码如下所示：

```
[root@master ~]# hdfs NameNode -format
```

11. HDFS启动

HDFS启动命令如下所示：

```
[root@master ~]# start-dfs.sh
```

在主机浏览器中输入http://192.168.137.100:50070/可以查看启动情况。在导航栏中点击“Datanodes”选项，可以查看集群成功配置了两个DataNodes节点，如图2-48所示。

Hadoop Overview Datanodes Datanode Volume Failures Snapshot Startup Progress Utilities

Datanode Information

In operation

Node	Last contact	Admin State	Capacity	Used	Non DFS Used	Remaining	Blocks	Block pool used
master:50010 (192.168.137.102:50010)	2	In Service	16.99 GB	4 KB	2.3 GB	14.69 GB	0	4 KB (0%)
slave1:50010 (192.168.137.101:50010)	2	In Service	16.99 GB	4 KB	2.29 GB	14.69 GB	0	4 KB (0%)

图2-48 集群成功配置了两个DataNodes节点

12. YARN启动

YARN启动命令如下所示：

```
[root@localhost ~]# start-yarn.sh
```

在地址栏中输入http://192.168.137.100:8088访问YARN可视化管理界面，如图2-49所示。

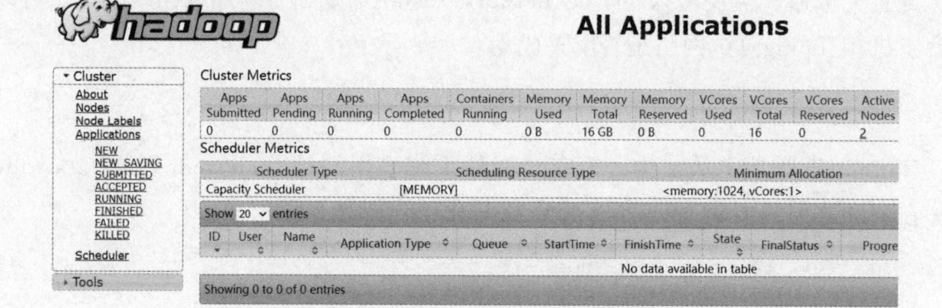

图2-49 YARN可视化管理界面

2.4 MapReduce开发环境的搭建

2.4.1 安装JDK

微课：
v2-9 Java环境遍历

1. 安装Java JDK软件包

JDK的安装界面如图2-50所示。

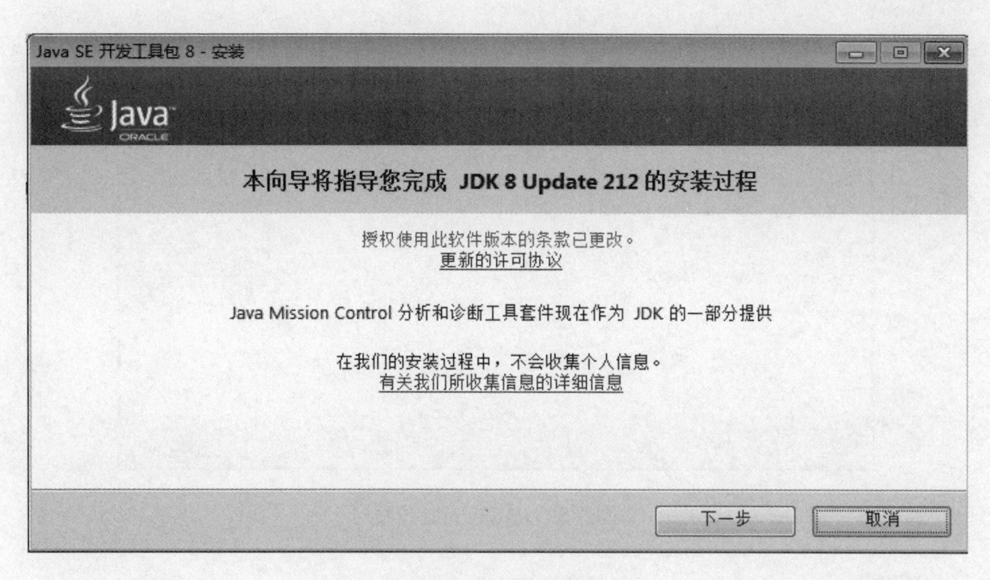

图2-50 JDK的安装界面

安装JDK过程中直接默认点击“下一步”按钮即可，如图2-51所示。

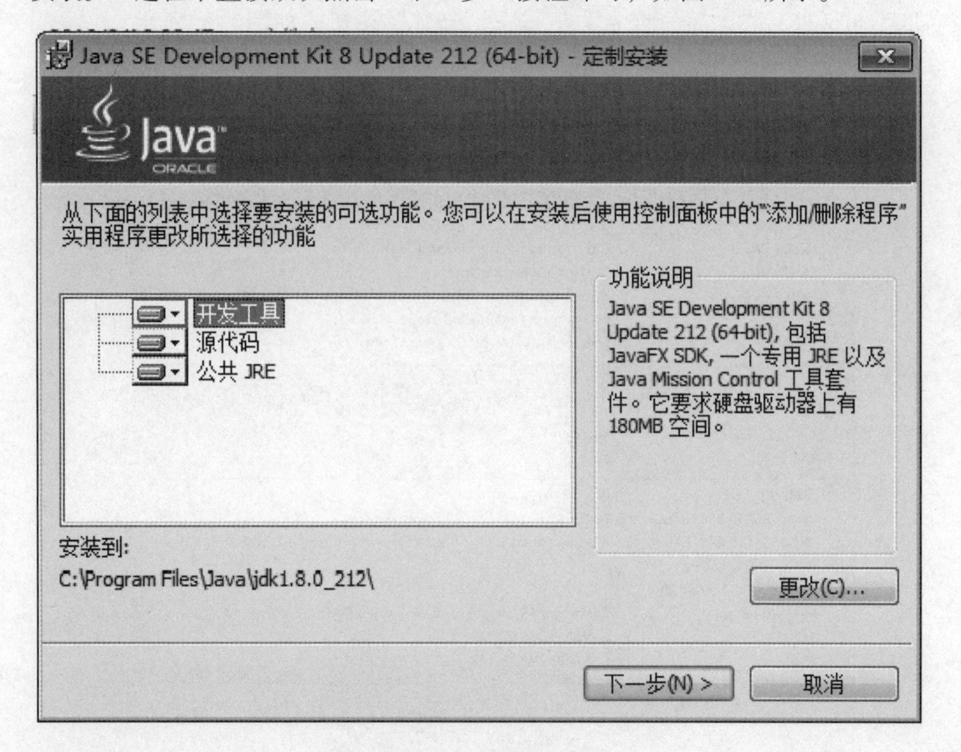

图2-51 安装JDK

2. 安装JRE

JRE的安装界面如图2-52所示。

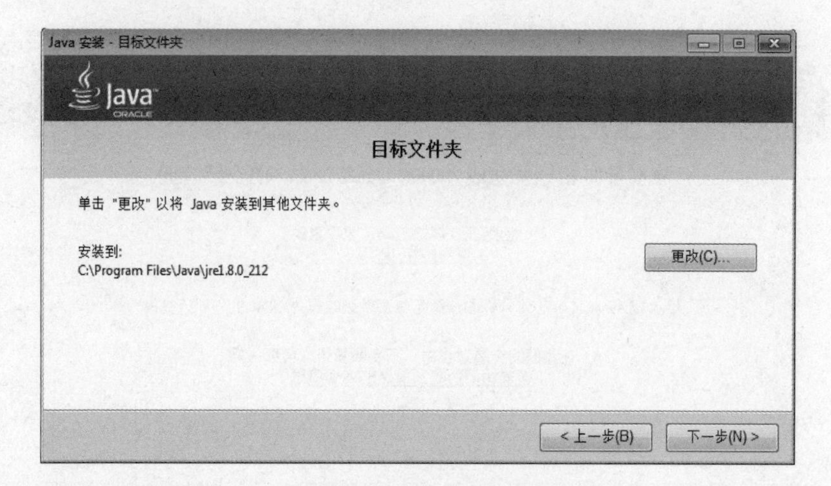

图2-52 JRE的安装界面

3.配置环境变量

右击“我的电脑”→“属性”→“高级系统设置”→“环境变量”，出现“环境变量”对话框，在“系统变量”下点击“新建”按钮，如图2-53所示。

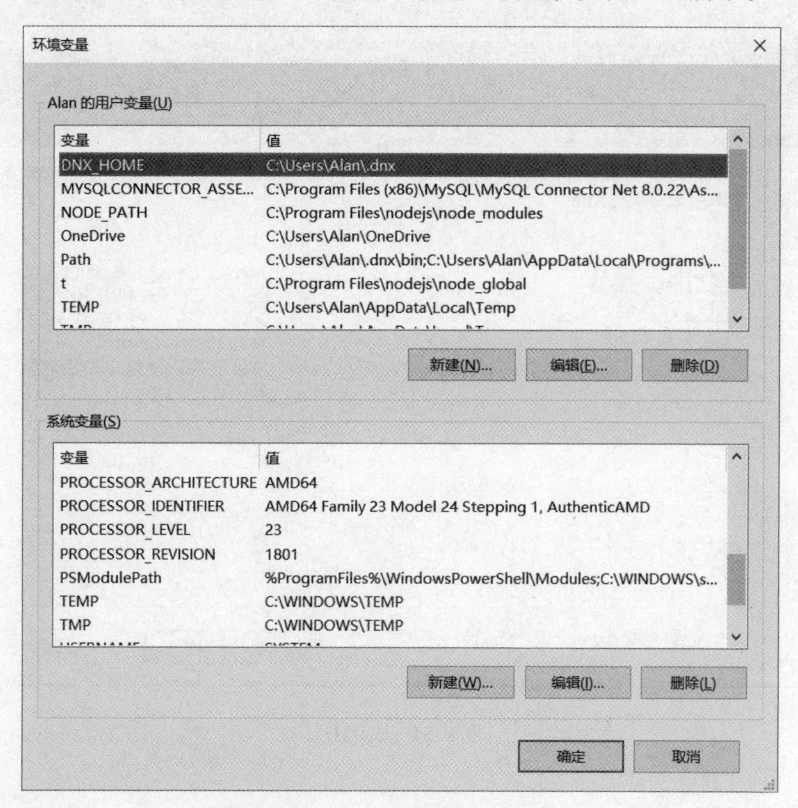

图2-53 “环境变量”对话框

在弹出来的“新建系统变量”对话框中，在“变量名”后面的文本框中输入“JAVA_HOME”，在“变量值”后面的文本框中输入对应的Java安装目录“C:\

Program Files\Java\jdk1.8.0_212”，点击“确定”按钮保存，如图2-54所示。

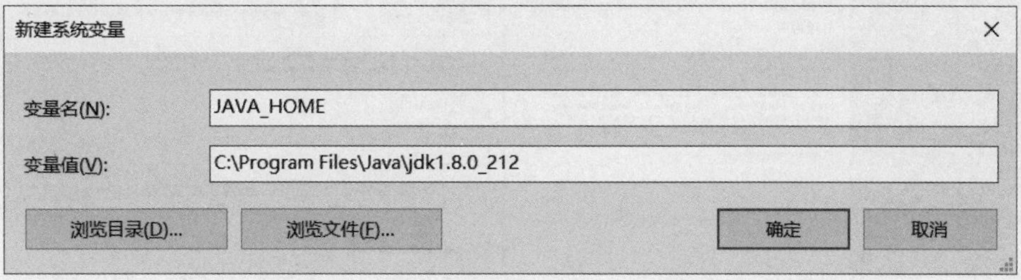

图2-54　“新建系统变量”对话框

在系统变量中找到Path，然后双击添加新的环境变量，在文末输入“%JAVA_HOME%\bin”、“%JAVA_HOME%\jre\bin”，如图2-55所示。

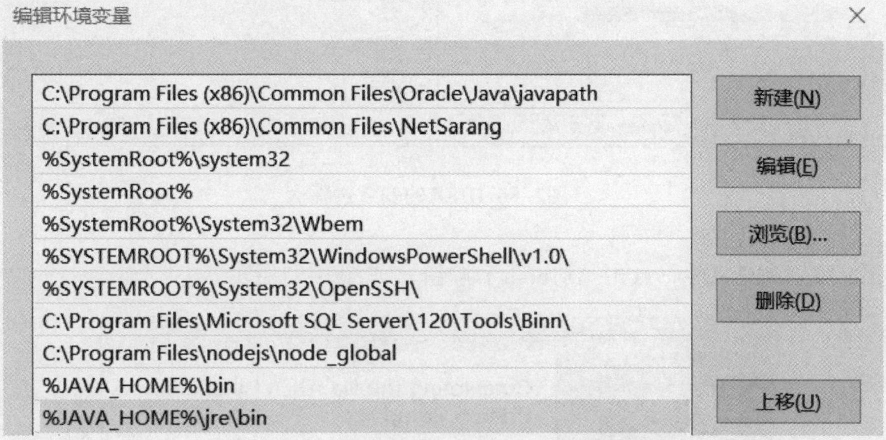

图2-55　编辑Path

新建完成后，验证环境变量是否正确。在cmd窗口中输入java -version，代码如下所示：

```
C:\Users\Alan>java -version
java version "1.8.0_212"
Java(TM) SE Runtime Environment (build 1.8.0_212)
Java HotSpot(TM) 64-Bit Server VM (build 25.181-b13, mixed mode)
```

2.4.2 安装IDEA

（1）双击打开“IntelliJ IDEA Community Edition Setup”安装包，安装过程中点击“Next”按钮直至安装完成，如图2-56所示。

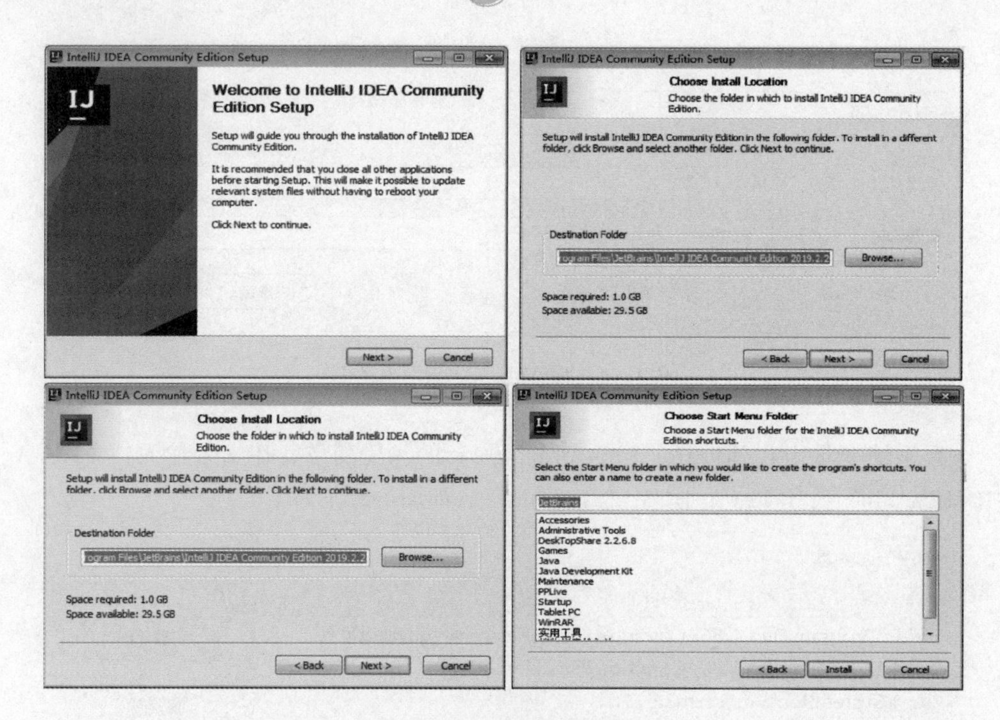

图2-56 IDEA的安装过程

（2）完成安装后，点击“Finish”按钮，如图2-57所示。

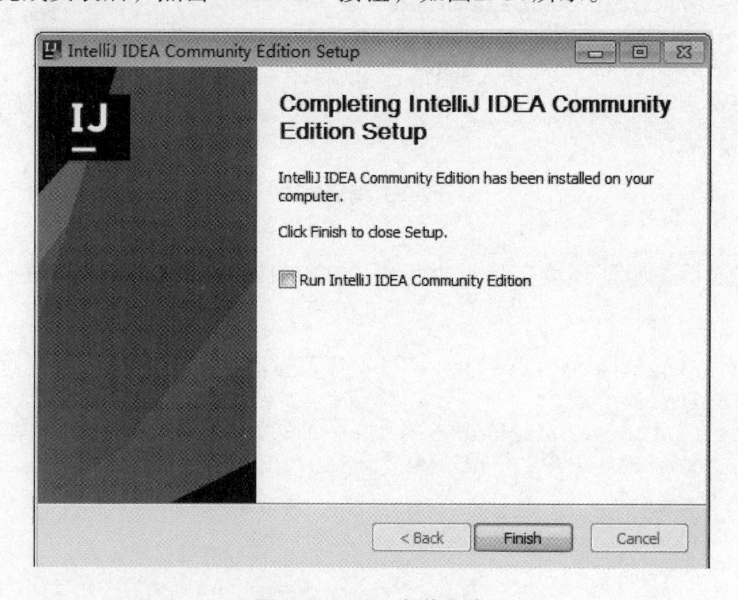

图2-57 IDEA安装完成

2.4.3 配置IDEA及新建测试项目

1. 新建项目

微课：v2-10 环境测试

点击“+Create New Project”，在弹出的窗口中选择“Java”，再点击“Next”按钮继续操作，如图2-58所示。

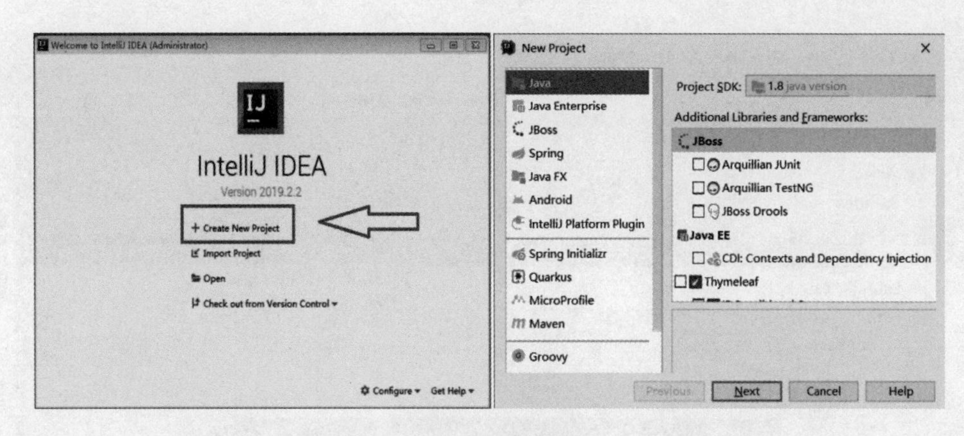

图2-58 创建项目

在“New Project”界面的“Project name”文本框中输入“demo”，再点击“Finish”按钮完成创建，如图2-59所示。

New Project
Project name: demo
Project location: D:\Hadoop\programe\demo
More Settings
Previous Finish Cancel Help

图2-59 为项目命名

2. 编写测试代码

为测试JDK、IDEA是否安装成功，在“demo”项目的“src”文件夹中右击，选择“New”→“Java Class”，将“Class”命名为“Test”，按回车键确认，如图2-60所示。

（1）代码如下所示：

```
public class Test {
    public static void main(String[] args) {
        System.out.println("好好学习，天天向上");
    }
}
```

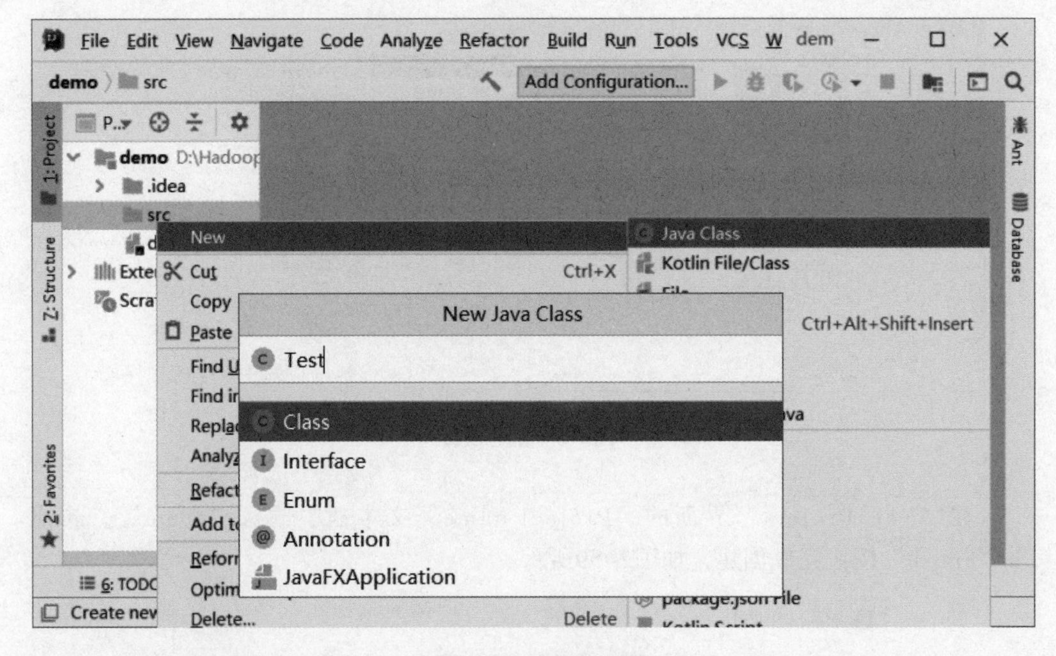

图2-60 创建源代码

（2）运行代码。

在代码空白处右击，选择“Run”运行代码，在IDEA控制台查看输出结果，如图2-61所示。

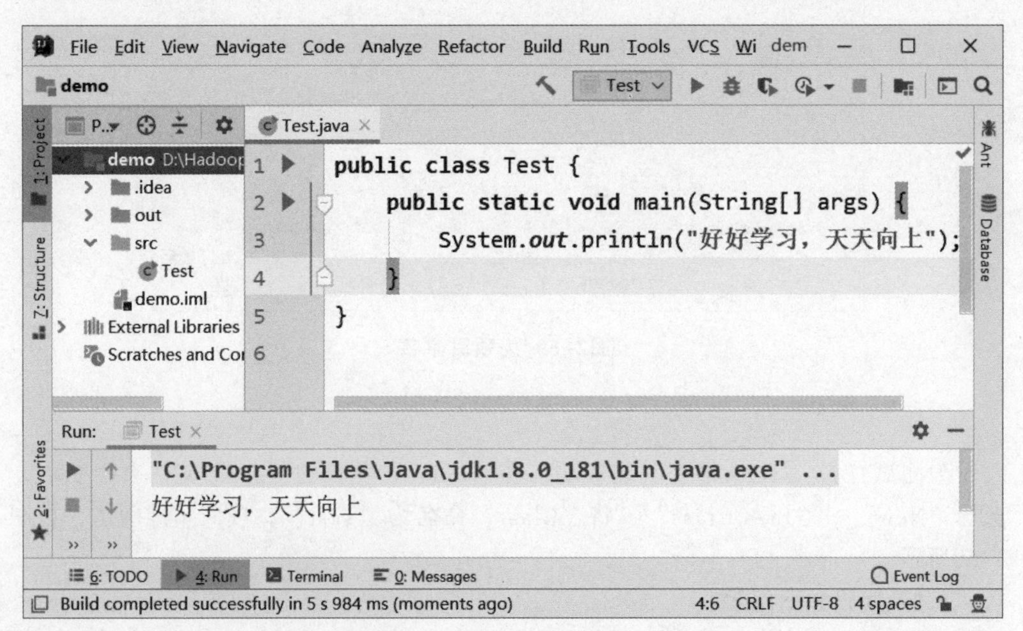

图2-61 代码运行后的成功界面

2.5 本章小结

本章主要讲解Hadoop集群的搭建过程。首先介绍虚拟机的安装、CentOS的安装，以及常用指令的使用；然后介绍网络配置与SSH设置；接着分析了Hadoop核心组件及其含义；最后讲解Hadoop集群的三种模式及其搭建方法，并讲解了IDEA的安装过程。

2.6 课后习题

1.选择题

（1）为了将当前目录的归档文件myfp.tgz解压缩到/tmp目录下，我们可以使用（　）。

A.tar xvjf myftp.tgz –C/tmp　　B.tar xvzf myftp.tgz –R/tmp

C.tar vzf myftp.tgz –X/tmp　　D.tar xvzf myftp.tgz/tmp

（2）以下（　）不属于Hadoop可以运行的模式。

A.单机（本地）模式　　B.伪分布式模式

C.互联模式　　D.分布式模式

（3）关于Hadoop单机模式和伪分布式模式的说法，正确的是（　）。

A.两者都起守护进程的作用，且守护进程运行在一台机器上

B.单机模式不使用HDFS，但加载守护进程

C.两者都不与守护进程交互，避免复杂性

D.后者比前者增加了HDFS输入/输出以及可检查内存使用的情况

（4）下列（　）程序通常与NameNode在同一个节点启动。

A.TaskTracker　　B.DataNode

C.SecondaryNameNode　　D.Jobtracker

（5）Hadoop是Java开发的，所以MapReduce只支持Java语言编写。（　）

A.正确　　B.错误

2.实操题

如果集群需要再添加一台服务器，需要如何操作。

第3章 Hadoop应用开发

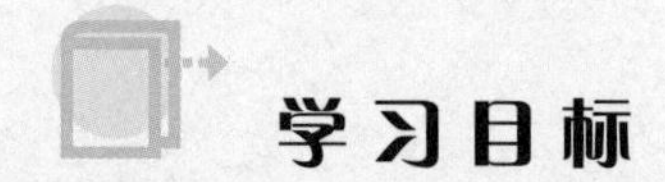

学习目标

（1）掌握Windows环境下MapReduce程序调试的方法。
（2）掌握IDEA创建项目的过程。
（3）掌握HDFS的常用shell指令用法。
（4）理解MapReduce程序执行的过程。
（5）掌握MapReduce程序设计方法。

思政目标

（1）了解应用的发展历史，理解行业职业规范和职业素养。
（2）培养团队精神和团队协作能力。

Hadoop集群可以安装在各个操作系统中。实际应用中，CentOS 7是使用广泛的操作系统，本书选择在CentOS 7上搭建Hadoop集群。为了方便教学，本章将在VMware Workstation软件中创建虚拟机，安装CentOS 7系统，并搭建集群。

提 示

hdfs指令练习可以使用伪分布式Hadoop。执行文件存储操作时，伪分布式Hadoop的速度要快于全分布式Hadoop的速度。

3.1 使用HDFS的shell指令

微课：
v3-1 hdfs
指令

HDFS是分布式文件系统，那么HDFS的操作，就是文件系统的基本操作，比如文件的创建、修改、删除、修改权限等，文件夹的创建、删除、重命名等。HDFS的操作命令类似于Linux的shell命令对文件的操作，如ls、mkdir、rm等。

（1）查看命令行帮助信息，代码如下：

```
[root@localhost ~]# hdfs dfs -help
```

（2）查看根目录下的文件与目录列表，代码如下：

```
[root@localhost ~]# hdfs dfs -ls /
```

（3）在根目录下创建文件夹，代码如下：

```
[root@localhost ~]# hdfs dfs -mkdir /test
```

通过操作可以发现，创建文件夹的指令与CentOS的操作指令类似，仅在代码前增加了“hdfs dfs”部分。

创建完成后，可以使用（2）中的代码查看，也可以通过可视化用户界面查看。在浏览器中输入192.168.137:50070，在出现的界面选择“Utilities”→“Browse the file system”查看文件、目录列表，如图3-1所示。在跳转的页面可以看到创建的“test”目录，如图3-2所示。

Hadoop　Overview　Datanodes　Datanode Volume Failures　Snapshot　Startup Progress　Utilities

Browse the file system
Logs

Overview 'localhost:9000' (active)

Started:	Fri Mar 18 21:00:44 EDT 2022
Version:	2.7.7, rc1aad84bd27cd79c3d1a7dd58202a8c3ee1ed3ac
Compiled:	2018-07-18T22:47Z by stevel from branch-2.7.7
Cluster ID:	CID-9a90aefe-6b3e-4003-9595-ed0d10c919dd
Block Pool ID:	BP-1675166625-192.168.137.140-1647430324188

图3-1　通过可视化用户界面查看文件、目录列表

Hadoop　Overview　Datanodes　Snapshot　Startup Progress　Utilities

Browse Directory

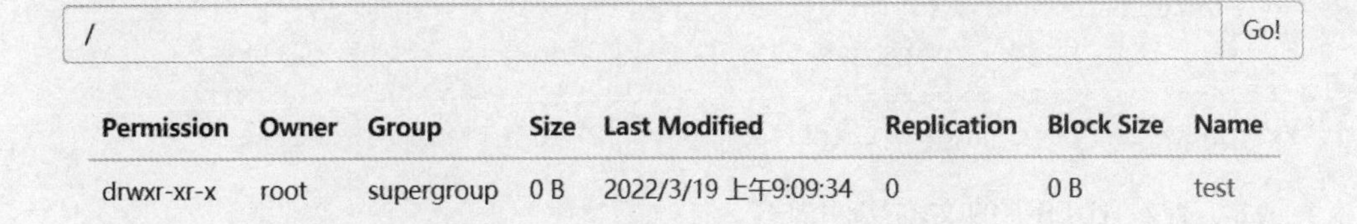

图3-2　创建的“test”目录

（4）上传文件到指定文件夹。

在CentOS主机的/opt目录中创建1.txt，输入内容后保存，并使用hdfs指令上传到集群的/test目录下，代码如下所示：

```
[root@localhost ~]# vi /opt/1.txt
富强 民主 文明 和谐 自由 平等 公正 法治 爱国 敬业 诚信 友善

[root@localhost ~]# hdfs dfs -put/opt/1.txt /test
[root@localhost ~]# hdfs dfs -ls /test
Found 1 items
-rw-r--r--   1 root supergroup        0 2022-03-18 21:15 /test/1.txt
```

通过（2）中的代码可以查看上传情况。

（5）从HDFS下载到本地。

将（4）上传的/test/1.txt下载到/root文件夹，代码如下所示：

```
[root@localhost ~] #hdfs dfs -get /test/1.txt /root
[root@localhost ~] #ls /root/
1.txt anaconda-ks.cfg
```

（6）复制文件。

将/test/1.txt复制到同目录下并重命名为2.txt，代码如下所示：

```
[root@localhost ~] #hdfs dfs -cp /test/1.txt /test/2.txt
[root@localhost ~] #hdfs dfs -ls /test
Found 2 items
-rw-r--r--   1 root supergroup        0 2022-03-18 21:15 /test/1.txt
-rw-r--r--   1 root supergroup        0 2022-03-18 21:33 /test/2.txt
```

将/test/2.txt复制到根目录下，代码如下所示：

```
[root@localhost ~] #hdfs dfs -cp /test/2.txt/
[root@localhost ~] #hdfs dfs -ls /
Found 2 items
-rw-r--r--   1 root supergroup        0 2022-03-18 21:34 /2.txt
drwxr-xr-x   - root supergroup        0 2022-03-18 21:33 /test
```

（7）移动文件。

在HDFS中，将/test/1.txt移动到根目录下，代码如下所示：

```
[root@localhost ~] #hdfs dfs -mv /test/1.txt/
[root@localhost ~] #hdfs dfs -ls/
Found 3 items
-rw-r--r--   1 root supergroup        0 2022-03-18 21:15 /1.txt
-rw-r--r--   1 root supergroup        0 2022-03-18 21:34 /2.txt
drwxr-xr-x   - root supergroup        0 2022-03-18 21:36 /test
```

（8）查看HDFS中的文本文件内容。

查看HDFS中的/test/1.txt内容，代码如下所示：

```
[root@localhost ~]# hdfs dfs -cat  /test/1.txt
富强 民主 文明 和谐 自由 平等 公正 法治 爱国 敬业 诚信 友善
```

（9）删除HDFS文件。

在HDFS中，某些文件不需要时，可通过删除命令进行操作。其中第二行代码中的“-r”参数用于删除文件和目录。

```
[root@localhost ~]# hdfs dfs -rm /test/1.txt
[root@localhost ~]# hdfs dfs -rm -r /test
```

3.2 使用Java操作HDFS

微课：
v3-2 HDFS开发环境配置

除了使用shell指令操作HDFS外，还可以通过编写Java程序操作HDFS。Java程序通过Hadoop提供的文件操作类执行读/写HDFS文件、上传HDFS文件等操作。这些文件操作类都在org.apache.hadoop.fs包中（详见Hadoop官方网站的Java API文档）。

API（Application Programming Interface，应用程序编程接口）是一些预先定义的函数。API使应用程序与开发人员获得了一种重要功能:无须访问源码或理解内部工作细节，即可访问一组例程。

3.2.1 导入Hadoop开发包

1. 导入准备

解压hadoop-2.7.7安装包，选取以下几个jar包并保存到一个目录下。

（1）在hadoop-2.7.7\share\hadoop\common\下的三个jar包。

（2）在hadoop-2.7.7\share\hadoop\common\lib\下的所有jar包。

（3）在hadoop-2.7.7\share\hadoop\hdfs\下的三个jar包。

2. 导入步骤

（1）打开demo项目，选择“File”→“Project Structure”，或者使用快捷键Ctrl+Alt+Shift+S，如图3-3所示。

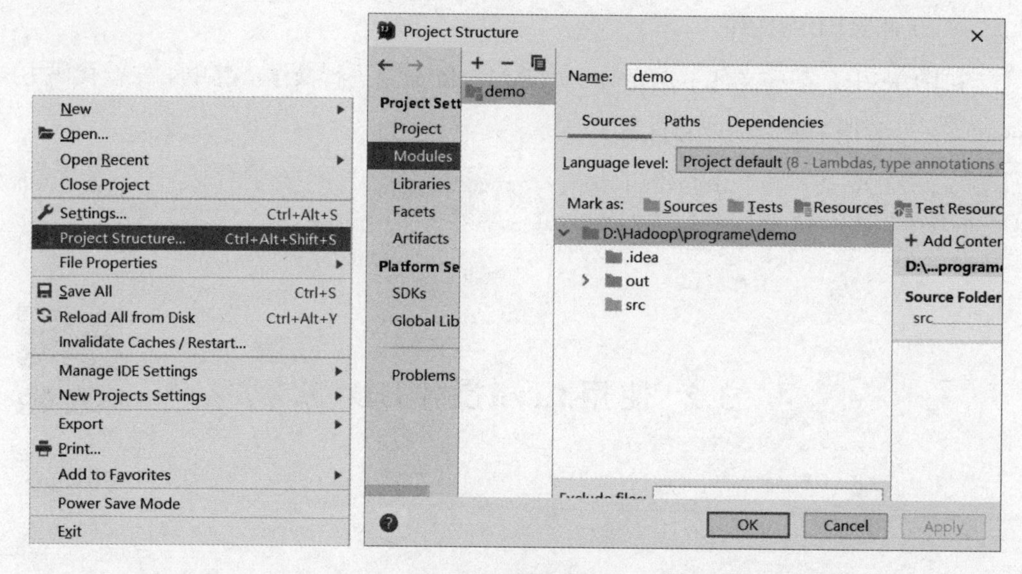

图3-3 查看依赖项

（2）点击“Dependencies”中的“+”选择合适的选项，如图3-4所示。

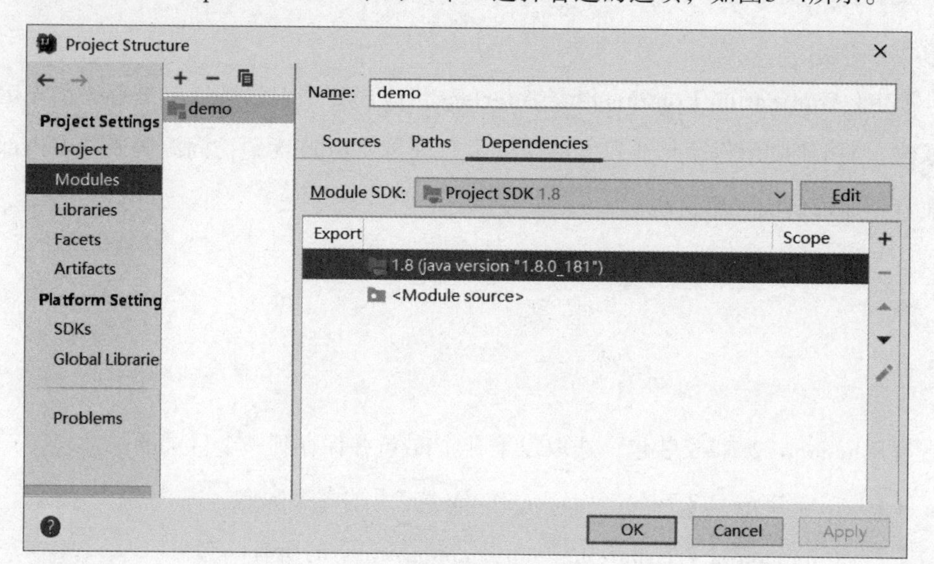

图3-4 添加依赖项

（3）选中从hadoop-2.7.7安装包中抽取出的jar包，在返回的界面中勾选jar包文件夹，如图3-5所示。

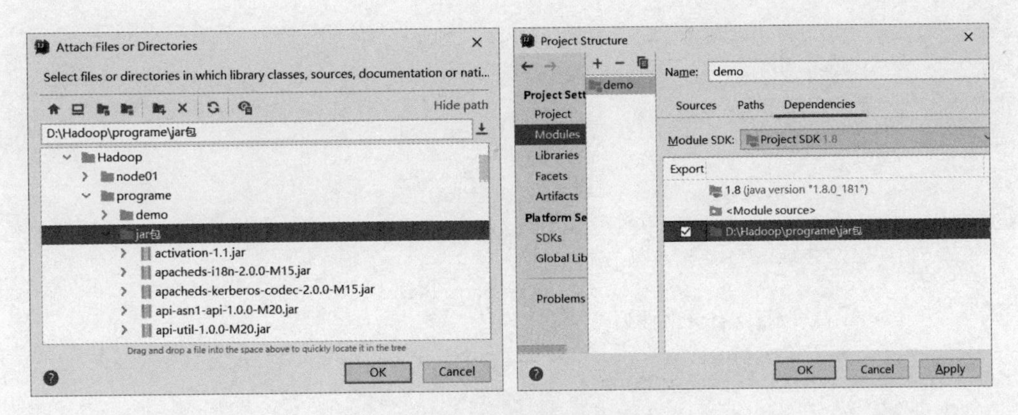

图3-5　选中依赖的jar包文件夹

（4）添加好项目的jar包之后，可以在项目文件夹中查看刚加入的jar包，如图3-6所示。

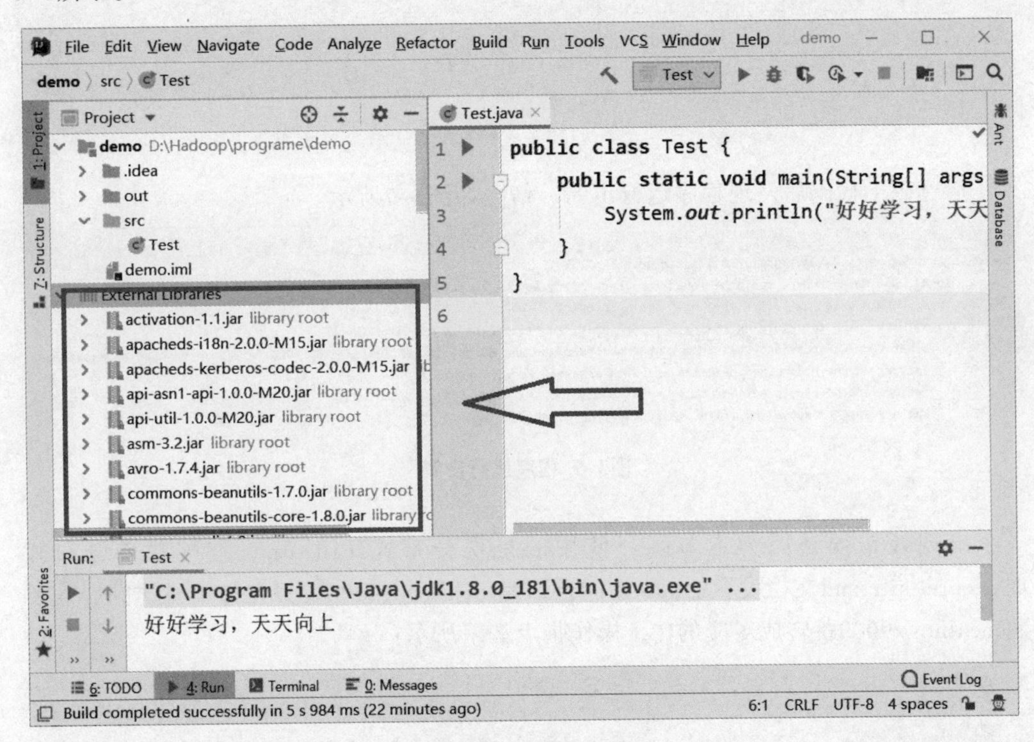

图3-6　导入依赖项列表

3.2.2 HDFS文件列表

在CentOS 7虚拟机中新建1.txt，再上传到hdfs根目录，并在hdfs中创建/test目录。打开Test.java，编写如下代码：

```
import org.apache.hadoop.conf.Configuration;
import org.apache.hadoop.fs.FileStatus;
import org.apache.hadoop.fs.FileSystem;
import org.apache.hadoop.fs.Path;
```

```
import org.apache.hadoop.fs.RemoteIterator;
import java.net.URI;
public class Test {
    public static void main(String[] args) {
        try {
            Configuration conf = new Configuration();
            URI uri = new URI("hdfs://192.168.137.100:9000");
            //生成hdfs连接对象
            FileSystem fs = FileSystem.get(uri,conf);
            //获取指定目录文件列表
            FileStatus[] paths = fs.listStatus(new Path("/"));
            //遍历访问返回的文件路径
            for(int i = 0;i<paths.length;i++) {
                System.out.println(paths[i].toString());
            }
            //关闭连接对象
            fs.close();
        } catch (Exception e) {
            e.printStackTrace();
        }
    }
}
```

运行以上代码后发现程序运行出错，信息如图3-7所示。

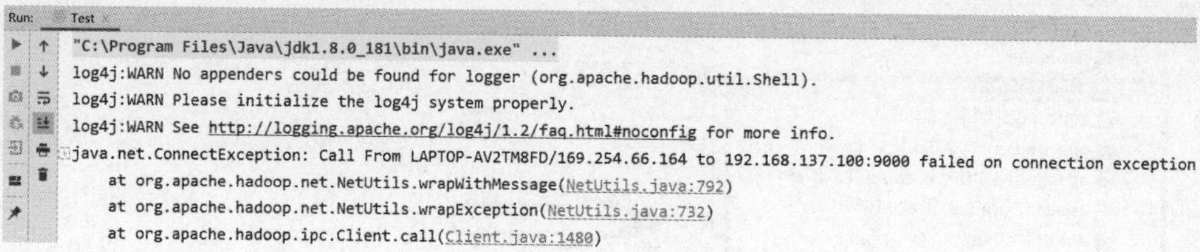

```
Run:  Test
"C:\Program Files\Java\jdk1.8.0_181\bin\java.exe" ...
log4j:WARN No appenders could be found for logger (org.apache.hadoop.util.Shell).
log4j:WARN Please initialize the log4j system properly.
log4j:WARN See http://logging.apache.org/log4j/1.2/faq.html#noconfig for more info.
java.net.ConnectException: Call From LAPTOP-AV2TM8FD/169.254.66.164 to 192.168.137.100:9000 failed on connection exception
    at org.apache.hadoop.net.NetUtils.wrapWithMessage(NetUtils.java:792)
    at org.apache.hadoop.net.NetUtils.wrapException(NetUtils.java:732)
    at org.apache.hadoop.ipc.Client.call(Client.java:1480)
```

图3-7 程序运行出错

出现该错误是因为本节练习使用的是伪分布式Hadoop，在配置伪分布式时，core-site.xml文档设置的hdfs访问接口是localhost:9000，只允许本机访问，需要将localhost:9000跳转成对应的IP，操作指令如下所示：

```
[root@localhost ~]# vi $HADOOP_HOME/etc/hadoop/core-site.xml

<configuration>
    <property>
        <name>fs.defaultFS</name>
        <value>hdfs://192.168.137:9000</value>
    </property>
    <property>
        <name>hadoop.tmp.dir</name>
        <value>/usr/local/hadoop-2.7.7/tmp</value>
    </property>
</configuration>
```

修改完成后，重新启动集群。再运行代码，运行结果如图3-8所示。

```
Run: Test ×
"C:\Program Files\Java\jdk1.8.0_181\bin\java.exe" ...
log4j:WARN No appenders could be found for logger (org.apache.hadoop.util.Shell).
log4j:WARN Please initialize the log4j system properly.
log4j:WARN See http://logging.apache.org/log4j/1.2/faq.html#noconfig for more info.
FileStatus{path=hdfs://192.168.137.100:9000/1.txt; isDirectory=false; length=0; replication=1; blocks
FileStatus{path=hdfs://192.168.137.100:9000/test; isDirectory=true; modification_time=1647660758835;

Process finished with exit code 0
```

图3-8 程序运行结果

从运行结果看，编写的程序成功读取到hdfs中的文件列表。

3.2.3 HDFS上传文件

使用java.net.URI从Hadoop文件系统中读取文件信息，通过“FileSystem.copyFromLocalFile（Path src，Path dst）”可将本地文件上传到HDFS指定的目录中。本节首先在C盘根目录新建文件“ceshi.txt”，并在该文件中写入“富强 民主 文明 和谐 自由 平等 公正 法治 爱国 敬业 诚信 友善”；然后将文件上传到hdfs根目录。代码如下所示：

```
import org.apache.hadoop.conf.Configuration;
import org.apache.hadoop.fs.FileStatus;
import org.apache.hadoop.fs.FileSystem;
import org.apache.hadoop.fs.Path;
import org.apache.hadoop.fs.RemoteIterator;

import java.net.URI;

public class Test {
    public static void main(String[] args) {
        try {
            Configuration conf = new Configuration();
            URI uri = new URI("hdfs://192.168.137.100:9000");
            //生成hdfs连接对象
            FileSystem fs = FileSystem.get(uri,conf);
            //获取指定目录文件列表
            Path localPath = new Path("C://ceshi.txt");
            Path hdfsPath = new Path("/");
            //上传文件
            fs.copyFromLocalFile(localPath,hdfsPath);
            //关闭连接对象
            fs.close();
            System.out.print("成功上传文件！");
        }catch (Exception e){
            e.printStackTrace();
        }
    }
}
```

代码在运行时出现错误，如图3-9所示。

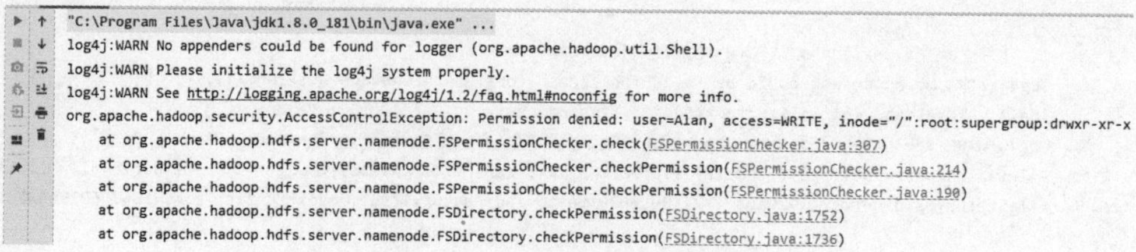

```
"C:\Program Files\Java\jdk1.8.0_181\bin\java.exe" ...
log4j:WARN No appenders could be found for logger (org.apache.hadoop.util.Shell).
log4j:WARN Please initialize the log4j system properly.
log4j:WARN See http://logging.apache.org/log4j/1.2/faq.html#noconfig for more info.
org.apache.hadoop.security.AccessControlException: Permission denied: user=Alan, access=WRITE, inode="/":root:supergroup:drwxr-xr-x
    at org.apache.hadoop.hdfs.server.namenode.FSPermissionChecker.check(FSPermissionChecker.java:307)
    at org.apache.hadoop.hdfs.server.namenode.FSPermissionChecker.checkPermission(FSPermissionChecker.java:214)
    at org.apache.hadoop.hdfs.server.namenode.FSPermissionChecker.checkPermission(FSPermissionChecker.java:190)
    at org.apache.hadoop.hdfs.server.namenode.FSDirectory.checkPermission(FSDirectory.java:1752)
    at org.apache.hadoop.hdfs.server.namenode.FSDirectory.checkPermission(FSDirectory.java:1736)
```

图3-9 代码在运行时出现错误

由于Hadoop文件系统操作的权限验证依靠Linux系统，读取文件时不会报错，但是在写入文件时会验证用户名。如果计算机主机的用户名不是root，则会被hdfs拒绝访问，需要修改hdfs-site.xml文件，将权限检查关闭即可运行程序，如下所示：

```
[root@localhost ~] # vi $HADOOP_HOME/etc/hadoop/hdfs-site.xml

<configuration>
<property>
        <name>dfs.replication</name>
        <value>1</value>
</property>
<property>
    <name>dfs.permissions.enabled</name>
    <value>false</value>
</property>
</configuration>
```

3.2.4 读取HDFS文件数据

使用java.net.URI对象打开一个数据流，从建立的数据连接中读取数据流（stream）来实现将HDFS远端文件下载到本地磁盘中。本节将第3.2.3节上传的“ceshi.txt”下载到“D盘”，程序代码如下，文件读取结果如图3-10所示。

```
import org.apache.hadoop.conf.Configuration;
import org.apache.hadoop.fs.*;

import java.net.URI;

public class Test {
    public static void main(String[] args) {
        try {
            Configuration conf = new Configuration();
            URI uri = new URI("hdfs://192.168.137.100:9000");
            //生成hdfs连接对象
            FileSystem fs = FileSystem.get(uri,conf);
            FSDataInputStream inputStream =
                fs.open(new Path("/ceshi.txt"));
            byte[] buf = new byte[1024];
            int len = 0;
            while((len = inputStream.read(buf)) != -1){
```

```
                System.out.println(new String(buf,0,len));
            }
            //关闭连接对象
            fs.close();
        }catch (Exception e){
            e.printStackTrace();
        }
    }
}
```

```
"C:\Program Files\Java\jdk1.8.0_181\bin\java.exe" ...
log4j:WARN No appenders could be found for logger (org.apache.hadoop.util.Shell).
log4j:WARN Please initialize the log4j system properly.
log4j:WARN See http://logging.apache.org/log4j/1.2/faq.html#noconfig for more info.
富强 民主 文明 和谐 自由 平等 公正 法治 爱国 敬业 诚信 友善
```

图3-10 文件读取结果

3.2.5 新建HDFS目录

使用java.net.URI类来打开一个数据连接，建立数据连接获取文件路径，通过FileSystem API获取HDFS数据，并在HDFS中创建文件或目录，代码如下所示：

```
import org.apache.hadoop.conf.Configuration;
import org.apache.hadoop.fs.*;
import java.net.URI;
public class Test {
    public static void main(String[] args) {
        try {
            Configuration conf = new Configuration();
            URI uri = new URI("hdfs://192.168.137.100:9000");
            //生成hdfs连接对象
            FileSystem fs = FileSystem.get(uri,conf);
            Path newDir = new Path ("newDir");
            if (!fs.exists(newDir)){
                fs.mkdirs(newDir);
                System.out.println("目录创建成功");
            } else {
                System.out.println("目录已存在");
            }
            //关闭连接对象
            fs.close();
        }catch (Exception e){
            e.printStackTrace();
        }
    }
}
```

3.2.6 删除HDFS文件、目录

使用java.net.URI对象打开一个数据连接，建立数据连接获取文件路径，通过”FileSystem.delete（Path　f，Boolean recursive）”可删除指定的HDFS文件，其中f为需要删除文件的完整路径，recursive用来确定是否进行递归删除。本节删除第3.2.5节创建的目录，代码如下所示：

```
import org.apache.hadoop.conf.Configuration;
import org.apache.hadoop.fs.*;
import java.net.URI;
public class Test {
    public static void main(String[] args) {
        try {
            Configuration conf = new Configuration();
            URI uri = new URI("hdfs://192.168.137.100:9000");
            //生成HDFS连接对象
            FileSystem fs = FileSystem.get(uri,conf);
            Path delPath = new Path("/newDir");
            if(fs.exists(delPath)) {
                fs.delete(delPath,true);
                System.out.println("删除成功");
            } else {
                System.out.println("删除失败");
            }
            //关闭连接对象
            fs.close();
        }catch (Exception e) {
            e.printStackTrace();
        }
    }
}}
```

3.3 认识MapReduce

3.3.1 MapReduce结构

HDFS和MapReduce实现是完全分离的，并不是没有HDFS就不能执行MapReduce运算。MapReduce是一个分布式运算程序的编程框架，是用户开发“基于Hadoop的数据分析应用”的核心框架。核心功能是将用户编写的业务逻辑代码和自带默认组件整合成一个完整的分布式运算程序，并发运行在Hadoop集群上。

MapReduce将整个并行计算过程抽象到两个函数。

Map（映射）：对由一些独立元素组成的列表的每一个元素执行指定的操作，可

以高度并行。

Reduce（归约）：归约过程，将若干组映射结果进行汇总并输出。

一个简单的MapReduce程序只需要指定Map()、Reduce()、Input和Output，剩下的事情由框架完成。

基于MapReduce编写出来的应用程序能够运行在大型集群上，并以一种可靠、容错的方式并行处理上T级别的数据集。一个Map/Reduce作业通常会将输入的数据集切分为若干个独立的数据块，由/Map任务（task）/以完全并行的方式处理它们。框架会对Map的输出先进行排序，然后将结果输入给/Reduce任务/。通常作业的输入和输出都会被存储在文件系统中。整个框架负责任务的调度和监控，本节主要分析MapReduce程序设计过程，MapReduce整体结构如图3-11所示。

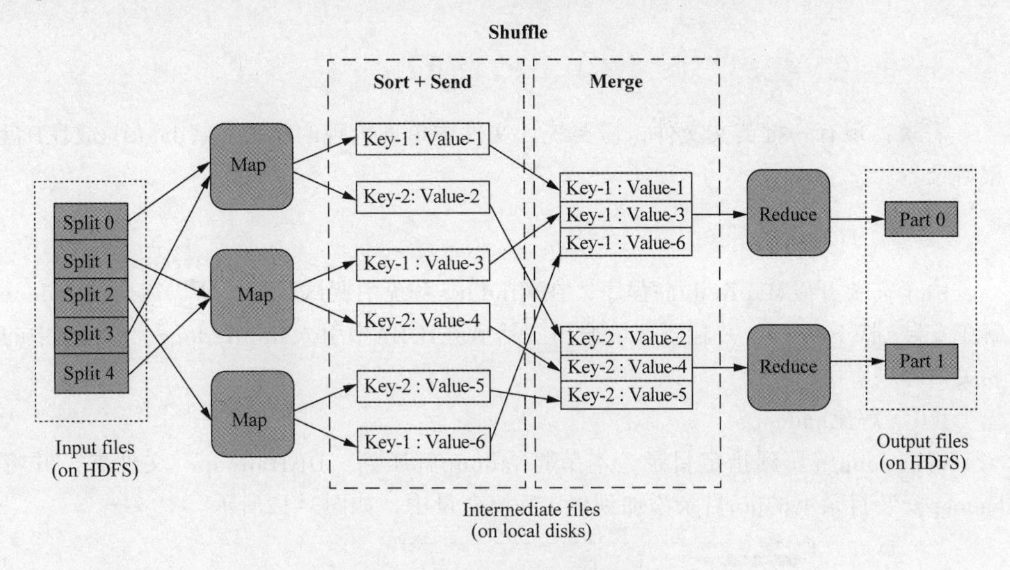

图3-11 MapReduce整体结构

在图3-11中，MapReduce程序分别由几个阶段构成，其含义如表3-1所示。

表3-1 MapReduce程序的构成

阶　段	内　容
Input	输入文件的存储位置，可以是HDFS文件位置，也可以是本地文件位置
Map	自己编写映射逻辑
Shuffle	是我们不需要编写的模块，但却是十分关键的模块。Shuffle阶段需要从所有Map主机上将相同的Key的Key-Value对组合在一起，传给Reduce主机，并进入Reduce()函数里
Reduce	自己编写合并逻辑
最后阶段	最终结果存储在HDFS中

3.3.2 MapReduce基本数据类型

MapReduce对一些基本数据类型进行了封装，表3-2对比了Hadoop数据类型与Java数据类型。

表3-2 数据类型对比

数据类型	Hadoop数据类型	Java数据类型
布尔型	BooleanWritable	boolean
整型	IntWritable	int
长整型	LongWritable	long
浮点型	FloatWritable	float
双精度浮点型	DoubleWritable	double
字节	ByteWritable	byte
字符串	Text	string
数组	ArrayWritable	array
Map	MapWritable	map

3.3.3 MapReduce案例：WordCount

需求：现有一批英文文件，需要统计文件中单词出现的次数。将data.txt放在D盘根目录。

1. 开发前准备

由于需要开发MapReduce程序，在Windows系统中调试运行，所以需要将hadoop解压安装到Windows，并配置环境遍历。然后在IDEA中引入MapReduce、YARN中的相关依赖项。

（1）解压hadoop。

将hadoop解压到指定目录。本节将hadoop解压到“D:\Hadoop”文件夹，并将hadoop安装目录下的bin目录添加到Path环境变量中，如图3-12所示。

微课：
v3-3
MapReduce+
本地开发环境

图3-12 将hadoop安装目录下的bin目录添加到Path环境变量

（2）将hadoop.dll和winutils.exe加入hadoop目录的bin目录下。

（3）将下列jar包引入项目中：

- hadoop-2.7.7\share\hadoop\mapreduce下的所有jar包。
- hadoop-2.7.7\share\hadoop\mapreduce\lib\下的所有jar包。
- hadoop-2.7.7\share\hadoop\yarn下的所有jar包。
- hadoop-2.7.7\share\hadoop\yarn\lib\下的所有jar包。
- hadoop-2.7.7\share\hadoop\common\下的三个jar包。
- hadoop-2.7.7\share\hadoop\ common\lib\下的所有jar包。
- hadoop-2.7.7\share\hadoop\hdfs\下的三个jar包导入。

2. 代码编写

测试MR运行环境的配置是否正确，通过demo案例来完成验证，代码如下所示：

```
import org.apache.hadoop.conf.Configuration;
import org.apache.hadoop.fs.Path;
import org.apache.hadoop.io.IntWritable;
import org.apache.hadoop.io.Text;
import org.apache.hadoop.mapreduce.Job;
import org.apache.hadoop.mapreduce.Mapper;
import org.apache.hadoop.mapreduce.Reducer;
import org.apache.hadoop.mapreduce.lib.input.FileInputFormat;
import org.apache.hadoop.mapreduce.lib.output.FileOutputFormat;
import org.apache.hadoop.util.GenericOptionsParser;
import java.io.IOException;
import java.util.StringTokenizer;
public class Test {
    public static class TokenizerMapper
            extends Mapper<Object,Text,Text,IntWritable> {

        private final static IntWritable one = new IntWritable(1);
        private Text word = new Text();

        public void map(Object key,Text value,Context context
        ) throws IOException,InterruptedException {
            StringTokenizer itr = new StringTokenizer(value.toString());
            while (itr.hasMoreTokens()) {
                word.set(itr.nextToken());
                context.write(word,one);
            }
        }
    }
    public static class IntSumReducer
            extends Reducer<Text,IntWritable,Text,IntWritable> {
        private IntWritable result = new IntWritable();

        public void reduce(Text key,Iterable<IntWritable> values,
            Context context
            ) throws IOException,InterruptedException {
                int sum = 0;
```

```
                for (IntWritable val:values) {
                    sum += val.get();
            }
            result.set(sum);
            context.write(key,result);
        }
    }

    public static void main(String[] args) throws Exception {
        Configuration conf = new Configuration();
        Job job = Job.getInstance(conf,"word count");
        job.setJarByClass(Test.class);
        job.setMapperClass(TokenizerMapper.class);
        job.setReducerClass(IntSumReducer.class);
        job.setOutputKeyClass(Text.class);
        job.setOutputValueClass(IntWritable.class);
        FileInputFormat.addInputPath(job, new Path("d://ceshi.txt"));
        FileOutputFormat.setOutputPath(job,new Path("D://result"));
        System.exit(job.waitForCompletion(true) ? 0:1);
    }
}
```

程序运行完毕后，可以看到D盘出现新的result文件夹。文件夹内的文件如图3-13所示。

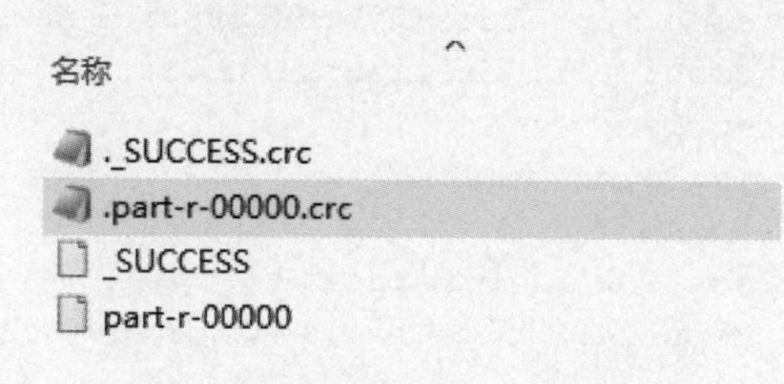

图3-13 WordCount运行文件结果

图3-13中的part-r-00000就是单词统计结果，选中文件，点击右键，选择打开方式，选择记事本，打开即可查看结果。

2. WordCount执行过程

WordCount的执行过程如图3-14所示。

1）Split

将程序的输入数据进行切分，每一个Split提交给一个MapTask执行。Split的数量可以自己定义。

2）Map

输入Split中的一个数据，对Split中的数据进行拆分，并以<key,value>对的格式保存数据，其中key的值为一个单词，value的值固定为1，如<I,1>、<wish,1>、……。

3）Combine/Shuffle/Sort

这几个过程在一些简单的MapReduce程序中并不需要我们关注，因为源代码中已经给出默认的Shuffle/Combine/Sort处理器，这几个过程的作用分别如下。

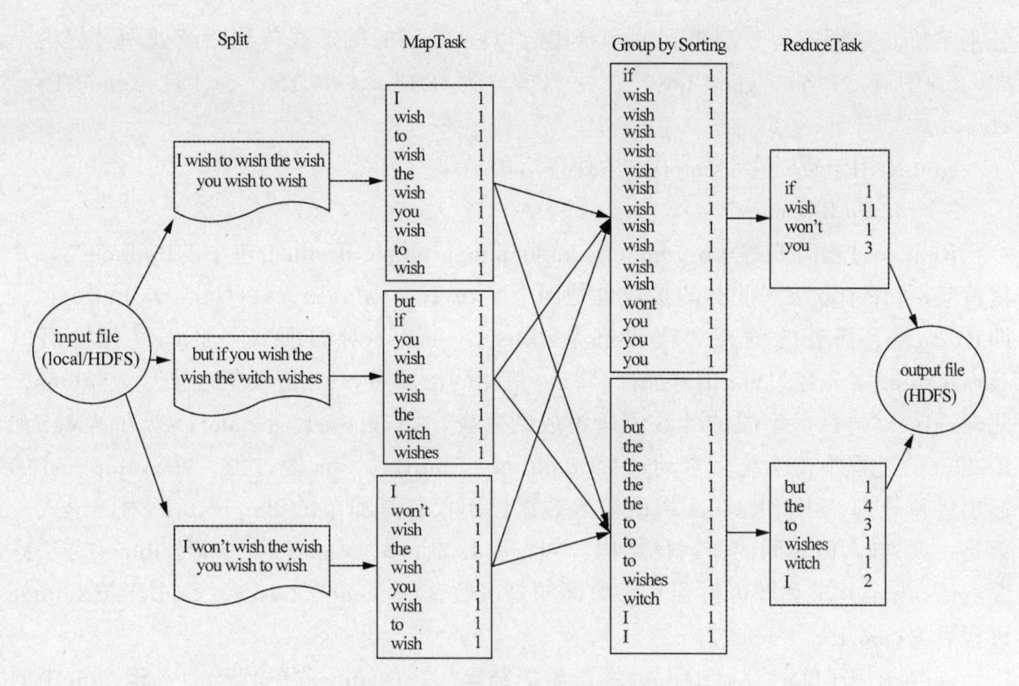

图3-14 WordCount的执行过程

（1）Combine：将MapTask产生的结果在本地节点上进行合并、统计等，以减少后续集群间Shuffle过程所要传输的数据量。

（2）Shuffle/Sort：将每个MapTask的处理结果在集群间进行传输、排序，并将其作为Reduce端的输入数据。

4）Reduce

Reduce Task的输入数据已经不是简单的<key,value>对，而是经过排序之后与key值相同的<key,value>对。ReduceTask会对其进行统计处理，产生最终的输出数据。

3. 代码分析

1）main()函数

从main()函数可知，程序通过job.getInstance()初始化一个job对象，再配置Mapper、Reducer需要调用的类，即绑定执行map、reduce过程需要调用的类。然后设定out过程中的<key,value>对类型，并定义输出。在FileInputFormat、FileOutputFormat中设置输入/输出路径。

2）TokenizerMapper

MapReduce程序需要继承org.apache.hadoop.mapreduce.Mapper包中的Reducer类，并在这个类的继承类中至少自定义实现map()方法。其中org.apache.hadoop.mapreduce.Mapper包要求的参数有四个（keyIn、valueIn、keyOut、valueOut），即map()任务的输入和输出都是<key，value>对的形式。

案例中，定义one作为基准数1，同时声明word字符类型对象，用于存储split过程中的单词。Hadoop加载Mapper类后，会调用map()函数。将文本中每一行传输到map中，因此map的value对象持有的是一行字符串。调用StringTokenizer对象用来分割字符。最后遍历分割得到的字符串，以key-value的形式传到后续处理过程中。如value中有“I am a good boy”，可以拆分为<I,1>、<am,1>、<a,1>、<good,1>、<boy,1>。

context用于暂时存储map()方法处理后的结果。

3）IntSumReducer

Reduce过程需要继承org.apache.hadoop.mapreduce.Reducer包中的Reduce类，并重写其reduce()方法。此类的参数也是四个（keyIn、valueIn、keyOut、valueOut），即Reduce()任务的输入和输出都是<key，value>对的形式。Map过程输出的<key,value>对先经过Shuffle过程，将key值相同的所有value值聚集起来形成values，此时values对应key字段的计数值所组成的列表，如<good,1>、<am,1>。在从Map到Reduce的过程中，系统会自动进行Combine、Shuffle、Sort等过程，并对MapTask的输出进行处理，因此Reduce端的输入数据已不仅仅是简单的<key,value>对的形式，而是一系列key值相同的序列化结构，如<good,<1,1>>、<am,1>。所以reduce()方法只要遍历values并求和即可得到某个单词的总次数，如<good,<1,1>>、<am,1>被Reduce过程后为<am,1>、<good,2>。

context用于临时存储Reduce端产生的结果。在Reduce端的代码中，将value中的值进行累加，所得结果就是对应key值的单词在文本中出现的词频。

3.4 本章小结

本章主要介绍了HDFS和MapReduce的开发案例。首先讲解了HDFS的shell指令，然后针对常用的操作编写了Java程序，最后深入分析了MapReduce的WordCount案例。

3.5 课后习题

1.选择题

（1）有关MapReduce的输入/输出，说法错误的是（　）。

A. 链接多个MapReduce作业时，序列文件是首选格式

B. FileInputFormat中实现的getSplits()可以将输入数据划分为分片、分片数目和大小任意定义

C. 想完全禁止输出，可以使用NullOutputFormat

D. 每个Reduce过程需将它的输出写入自己的文件中，输出不需要分片

（2）MapReduce的Shuffle过程中，（　）操作是最后做的。

A.溢写　　B.分区　　C.排序　　D.合并

（3）下面关于MapReduce的描述中，不正确的是（　）。

A.MapReduce程序必须包含Mapper和Reducer

B.MapReduce程序的MapTask可以任意指定

C.MapReduce程序的ReduceTask可以任意指定

D.MapReduce程序的默认数据读取组件是TextInputFormat

（4）MapReduce编程模型中，以下组件（　）不是最后执行的。

A.Mapper　　B.Partitioner

C.Reducer　　D.RecordReader

2.编程题

在本章的MapReduce案例中，统计包含A的单词个数。

第4章 Hive数据仓库开发

学习目标

（1）了解Hive数据仓库的基本知识，包括Hive环境搭建、在Linux下部署MySQL以及基本命令的使用方法。

（2）掌握Hive的体系架构，掌握Hive中内部表、外部表、分区表以及分桶表的区别与联系。

（3）掌握Hive的SQL语句的使用规则和语法规则、Hive的正确的建表语句和Hive的分隔符的确定等。

思政目标

（1）为全面贯彻实施国家的大数据发展战略，切实推进大数据产业的全局发展，同时让学生树立正确的人生观、价值观和世界观。

（2）通过对Hive基本知识的讲解，培养学生的爱国主义精神、人文科学素养和社会责任感。

4.1 Hive概述

4.1.1 Hive简介

Hive是一个构建在Hadoop上的数据仓库框架。最初，Hive由Facebook开发，后来移交由Apache软件基金会开发，并作为一个Apache开源项目。Hive提供以表格的方式来组织与管理HDFS上的数据，以类SQL的方式来操作表格里的数据。Hive的设计目的是能够以类SQL的方式查询存放在HDFS上的大规模数据集，不必开发专门的MapReduce应用。本质上，Hive相当于一个MapReduce和HDFS的翻译终端。当用户提交Hive脚本时，首先，Hive运行时环境会将这些脚本翻译成MapReduce和HDFS并提交给集群；其次，Hive运行时环境调用Hadoop命令行接口向Hadoop集群执行这些MapReduce和HDFS操作；最后，Hadoop集群逐步执行这些MapReduce和HDFS操作。整个执行过程可概括如下。

（1）用户编写HiveQL并向Hive运行时环境提交该HiveQL。

（2）Hive运行时环境将HiveQL翻译成MapReduce和HDFS。

（3）Hive运行时环境调用Hadoop命令行接口或程序接口，向Hadoop集群提交翻译后的HiveQL。

（4）Hadoop集群执行HiveQL翻译后的MapReduce-APP或HDFS-APP。

由上述执行过程可知，Hive的核心是其运行时环境，该环境可将类SQL语句编译成MapReduce。Hive构建在基于静态批处理的Hadoop之上，例如，联机事务处理（OLTP）。Hive查询操作过程严格遵守Hadoop MapReduce的作业执行模型，Hive将用户的HiveQL语句通过解释器转换为MapReduce作业提交到Hadoop集群上，Hadoop监控作业执行流程，然后返回作业执行结果给用户。Hive并非为联机事务处理而设计，Hive并不提供实时的查询和基于行级的数据更新操作。Hive的最佳使用场景是大数据集的批处理作业，例如，网络日志分析。Hive没有专门的数据存储格式，也没有为数据建立索引。用户可以非常自由地组织Hive中的表，只需要在创建表的时候告诉Hive中的列分隔符和行分隔符，Hive就可以解析数据。Hive中的所有数据都存储在HDFS中，Hive包含表（Table）、外部表（External Table）、分区表（Partition Table）和桶表（Bucket Table）等数据模型。Hive中的Table和数据库中的Table在概念上是类似的。

4.1.2 Hive的特点

Hive作为Hadoop的一个数据仓库工具，其具备自身的优点，总结如下。

（1）简单易用。

Hive脚本基于SQL表达式语法，兼容大部分SQL-92语义和部分SQL-2003扩展语义，能让熟悉关系型数据库（如MySQL、SQL Server等）的开发者快速进入角色。

（2）可扩展。

Hive基于Hadoop实现，可以自由地扩展集群的规模，一般情况下不需要重启服务。

（3）延展性。

Hive支持用户自定义函数，用户可以根据自己的需求来定义自己的函数，实现相关的功能。

（4）容错性。

Hadoop具备良好的容错性，即使某个节点出现问题，SQL仍可执行完成。

（5）支持海量数据。

Hive适用场景一般为海量数据的存储处理挖掘海量数据的离线分析等符合当前大数据的时代背景。

4.1.3 Hive体系结构

Hive体系结构分析如下。

（1）最顶层的是User Interface（用户接口）层，包括Web UI、Hive Command Line和HDInsight三种方式。其中，Web UI可以通过浏览器界面操作Hive，一般只能查看。Hive Command Line也就是Hive Command Line Interface，我们称为CLI(Command Line Interface)，即Shell操作，这是使用最多的，因为我们可以编写Shell脚本来运行HDInsight，这是一种云技术驱动的Hadoop发行版。使用HDInsight，我们可以在HDInsight上预加载Hive库。基于Linux的HDInsight，我们可以使用Hive客户端WebHCat和HiveServer2。基于Windows的HDInsight，我们可以使用Hive客户端和WebHCat。

（2）MetaStore，也就是元存储。Hive的元数据存储在元数据库中，Hive默认自带有元数据库Derby，我们也可以使用其他的如MySQL数据库来存储元数据信息。Hive中的元数据包括表的名字、表的列、分区及其属性、表的属性（是否为外部表等）、表的数据所在目录等。

（3）HiveQL Process Engine，即HiveQL处理引擎。HiveQL类似于SQL，用于查询MetaStore上的模式信息，是MapReduce程序的传统方法的替代品之一，代替在Java中编写MapReduce程序。我们可以为MapReduce作业编写一个查询程序并执行它。

（4）Hive Excution Engine，即Hive执行引擎。HiveQL Process Engine和MapReduce的连接部分是Hive ExcutionEngine，Hive执行引擎处理查询并生成与MapReduce相同的结果。

（5）HDFS或HBase Data Storage，这是因为Hive主要进行数据分析，分析的数据来源于HDFS或HBase。

从上面的体系结构可以看出，Hive其实就是利用HiveQL Process Engine和Hive Excution Engine将用户的SQL语句解析成对应的MapReduce程序而已。

经过上面的说明，我们知道Hive的工作和Hadoop之间密不可分，它的工作执行要素之一需要依赖HDFS（也可以是HBase），执行要素之二需要依赖MapReduce。为了更加深入地理解其内部的执行机制，我们看一下Hive的工作流程图，如图4-1所示。

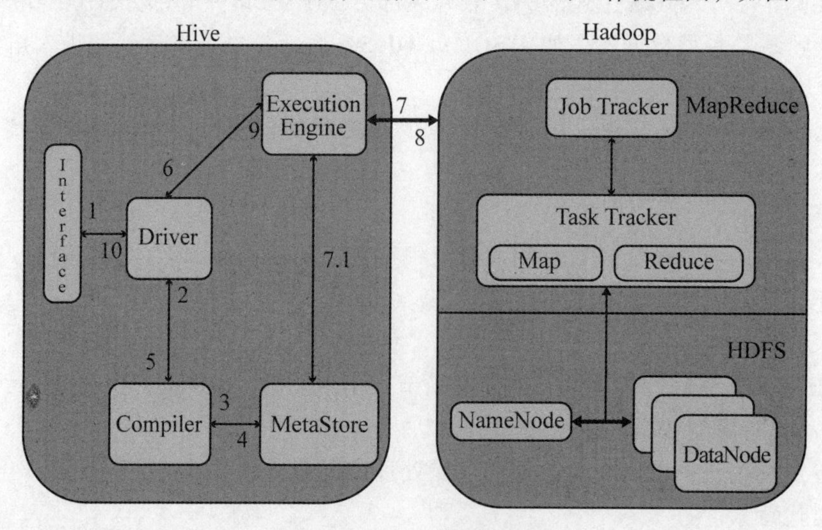

图4-1 Hive的工作流程图

图4-1展示了Hive工作时的内部流程，主要分为九步。

（1）执行查询：Hive接口比如Command Line会将用户输入的查询信息发送给数据库Driver（驱动），比如JDBC ODBC。

（2）Driver（驱动）在查询编译器的帮助下解析查询，并检查语法，生成一个查询计划。

（3）编译器向元数据库发送数据请求，然后元数据库返回数据给编译器。

（4）编译器检查查询计划是否满足各个条件和要素，然后将查询计划重新发送给数据库驱动，至此，解析和编译一个查询QL的工作已经完成。

（5）驱动就可以将查询计划发送给执行引擎，包含对应的数据信息。

（6）执行引擎处理我们的查询计划，会从元数据库中获取数据信息，然后将其转换为MapReduce Job，这就涉及了MapReduce的工作机制。首先这个查询计划会发送到JobTracker，然后分配给具体的TaskTracker去完成。MapReduce执行过程中会从HDFS读取数据。

（7）当执行引擎处理完查询计划后，会将结果返回到执行引擎处，这样执行引擎就能知道MapReduce执行结果的数据具体保存在HDFS的什么位置上。

（8）执行引擎处理完查询计划后，会将收到的结果返回给驱动。

（9）驱动将收到的结果返回给Hive接口，这样查询计划的执行者就可以看到查询计划的执行结果。

至此，相信你对Hive的工作流程有了一个清晰的认识。

4.1.4 Hive和普通关系型数据库的异同

Hive的基本语句和语法规则与普通关系型数据库的有相似之处，但也存在一些差异，总结如下。

- 查询语言。由于SQL被广泛应用在数据仓库中，因此，专门针对Hive的特性设计了类SQL的查询语言HQL。熟悉SQL开发的开发者可以很方便地使用Hive进行开发。
- 数据存储。Hive是建立在Hadoop之上的，所有Hive的数据都是存储在HDFS中的。而数据库则可以将数据保存在块设备或者本地文件系统中。
- 数据格式。Hive中没有定义专门的数据格式，数据格式可以由用户指定，用户定义数据格式需要指定三个属性：列分隔符（通常为空格、“\t”、“\x001”）、行分隔符（“\n”）以及读取文件数据的方法（Hive中默认有三个文件格式TextFile、SequenceFile以及RCFile）。在加载数据的过程中，由于不需要从用户数据格式到Hive定义的数据格式的转换，因此，Hive在加载的过程中不会对数据本身进行任何修改，而只是将数据内容复制或者移动到相应的HDFS目录中。在数据库中，不同的数据库有不同的存储引擎，定义了自己的数据格式。所有数据都会按照一定的组织存储，因此，数据库加载数据的过程会比较耗时。
- 数据更新。由于Hive是针对数据仓库应用而设计的，数据仓库的内容是读多写少，因此，Hive中不支持对数据的改写和添加，所有数据都是在加载时确定好的。而数据库中的数据需要经常修改，因此可以使用INSERT INTO…VALUES添加数据，使用UPDATE…SET修改数据。

- 索引。之前已经说过，Hive在加载数据的过程中不会对数据进行任何处理，甚至不会对数据进行扫描，因此也没有对数据中的某些Key建立索引。当Hive要访问数据中满足条件的特定值时，需要暴力扫描整个数据，因此访问延迟较高。由于MapReduce的引入，Hive可以并行访问数据，因此，即使没有索引，对于大数据量的访问，Hive仍然可以体现出优势。数据库中，通常会针对一个或者几个列建立索引，因此，对于少量的特定条件的数据访问，数据库可以有很高的效率、较低的延迟。由于数据的访问延迟较高，因此Hive不适合在线数据查询。
- 执行。Hive中大多数查询的执行是通过Hadoop提供的MapReduce来实现的（select * from tbl的查询不需要MapReduce）。而数据库通常有自己的执行引擎。
- 执行延迟。之前提到，Hive在查询数据的时候，由于没有索引，需要扫描整个表，因此延迟较高。另外一个导致Hive执行延迟高的因素是MapReduce框架。由于MapReduce本身具有较高的延迟，因此在利用MapReduce执行Hive查询时，也会有较高的延迟。相对地，数据库的执行延迟较低。当然，这个低是有条件的，即数据规模较小，当数据规模大到超过数据库的处理能力的时候，Hive的并行计算显然能体现出优势。
- 可扩展性。由于Hive是建立在Hadoop之上的，因此Hive的可扩展性与Hadoop的可扩展性一致（世界上最大的Hadoop集群在Yahoo!，2009年时，其规模约为4000个节点）。由于数据库受ACID语义的严格限制，因此其扩展性受到了限制。目前最先进的并行数据库Oracle在理论上的扩展能力也只有100台左右。
- 数据规模。由于Hive建立在集群上并可以利用MapReduce进行并行计算，因此它可以支持的数据规模很大；相应地，数据库可以支持的数据规模较小。

Hive和普通关系型数据库的不同点如表4-1所示。

表4-1 Hive和普通关系型数据库的不同点

区别	Hive	RDBMS
查询语言	HQL	SQL
数据存储	HDFS	Raw Device或Local FS
索引	无	有
执行	MapReduce	Executor
执行延迟	高	低
数据规模	大	小

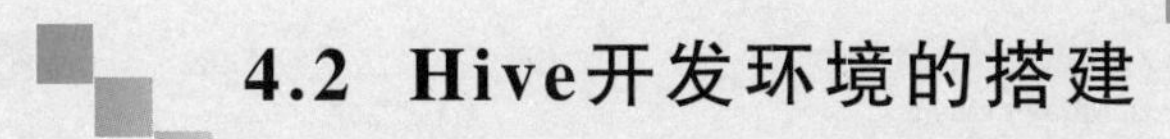

4.2 Hive开发环境的搭建

4.2.1 下载与安装Hive

在官网下载Hive安装包，下载地址为https://hive.apache.org/downloads.html，笔者下载的Hive版本为apache-hive-2.1.1-bin.tar。

在前面已经部署好的Zookeeper+Hadoop+HBase集群上安装Hive。Hive不需要以集群的方式提供服务，但是Hive的工作需要依赖HDFS，为了提高Hive的执行效率，我们决定将Hive安装到HDFS主节点上，HDFS主节点有两个，即huatec01和huatec02，它们彼此组成NameNode高可用。我们将Hive安装到其中一个节点即可，这里选择huatec01。Hive的安装十分简单，解压即可。

```
[root@huatec01 ~] #tar -xvf apache-hive-2.1.1-bin.tar -C /huatec/
```

我们将其安装到统一路径“/huatec”下。Hive的数据分为数据本身和元数据信息，其中数据本身保存在HDFS上，但是元数据信息需要保存到关系型数据库中。我们选用MySQL数据库作为元数据库，所以还需要安装MySQL。

4.2.2 安装元数据库

我们选用安装包的方式安装MySQL Server，因为Hive工作的时候需要依赖MySQL元数据库，所以在huatec01节点上安装MySQL，笔者采用yum仓库的方式进行安装，安装步骤如下所示。

1. 添加MySQL yum仓库

下载MySQL yum仓库，下载地址为https://dev.mysql.com/downloads/repo/yum/，如图4-2所示。

笔者的目标安装系统为CentOS 7，选择第一个进行下载，下载后获得一个rpm安装包，执行如下指令进行安装：

```
[root@huatec01 huatec] # yum localinstall mysql57-community-release-
    el7-11.
noarch.rpm
```

MySQL yum rpm包安装示意图如图4-3所示。

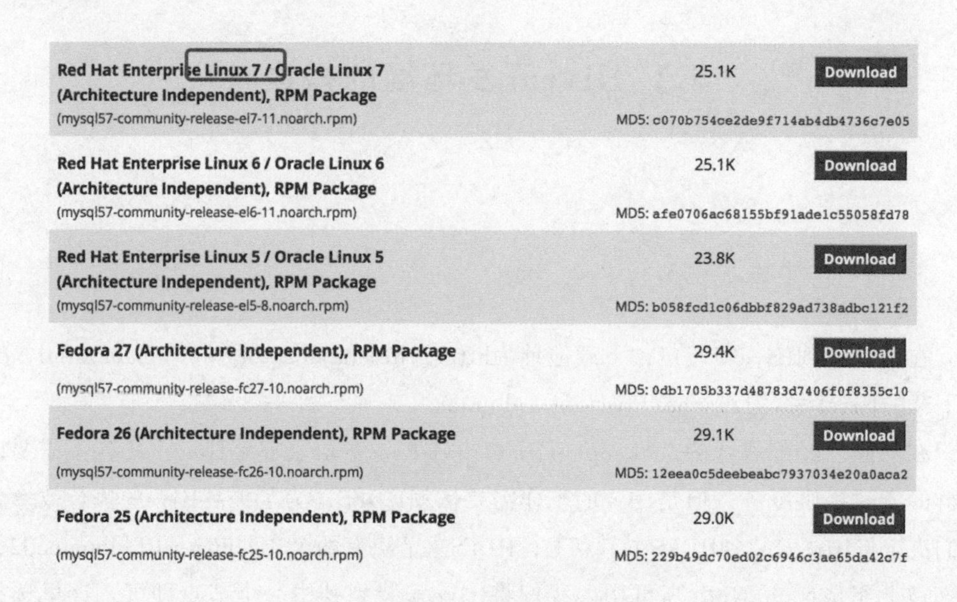

图4-2 MySQL yum仓库下载示意图

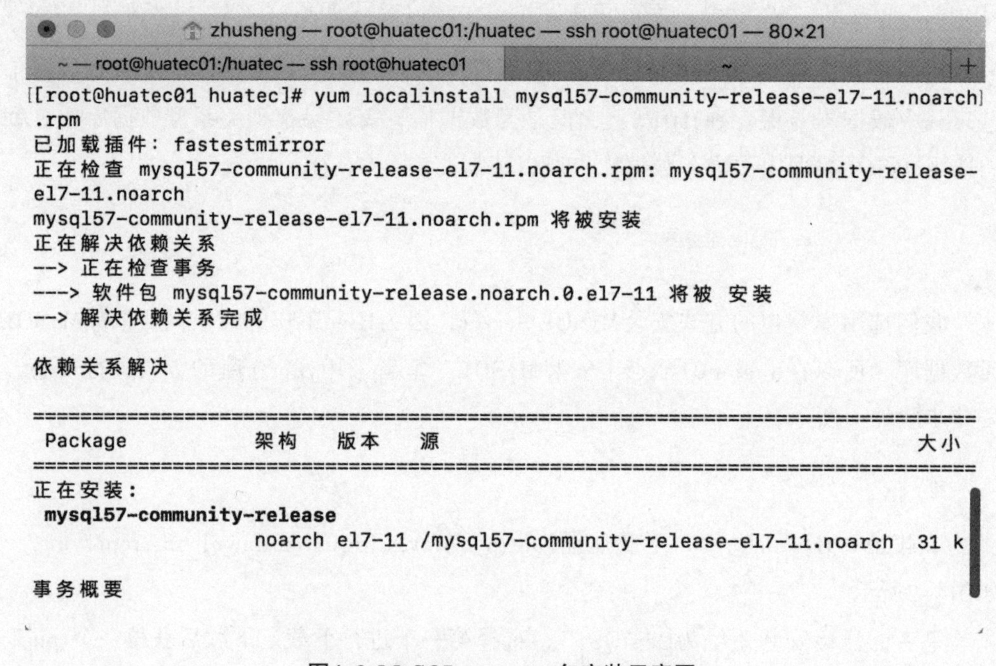

图4-3 MySQL yum rpm包安装示意图

上述指令会将MySQL yum repository添加到系统的yum repository列表中，并下载GnuPG key来检查软件安装包的完整性。

2. 让安装包可用

在上述操作中，我们向yum仓库添加了与MySQL有关的仓库，执行如下指令让MySQL有关的仓库可用，这样就可以使用yum方式来安装MySQL了。

```
[root@huatec01 yum] #yum repolist enabled | grep "mysql.-community.* "
```

在上述指令中，我们使用管道指令grep查看与MySQL社区版本相关的安装包。执行上述指令，让yum仓库可用的效果如图4-4所示。

```
[root@huatec01 yum]# yum repolist enabled | grep "mysql.*-community.*"
mysql-connectors-community/x86_64         MySQL Connectors Community        4
mysql-tools-community/x86_64              MySQL Tools Community             5
mysql57-community/x86_64                  MySQL 5.7 Community Server       22
[root@huatec01 yum]#
```

图4-4 让yum仓库可用的效果

3.安装MySQL

我们下载的MySQL pm包含了MySQL 5.5、5.6、5.7、7.5、7.6、8.0等诸多版本系列，默认安装的版本为5.7系列，下面使用yum指令进行查看，如图4-5所示。

```
[root@huatec01 yum]# yum repolist all |grep mysql
mysql-cluster-7.5-community/x86_64 MySQL Cluster 7.5 Community       禁用
mysql-cluster-7.5-community-source MySQL Cluster 7.5 Community - Sou 禁用
mysql-cluster-7.6-community/x86_64 MySQL Cluster 7.6 Community       禁用
mysql-cluster-7.6-community-source MySQL Cluster 7.6 Community - Sou 禁用
mysql-connectors-community/x86_64  MySQL Connectors Community        启用:
mysql-connectors-community-source  MySQL Connectors Community - Sour 禁用
mysql-tools-community/x86_64       MySQL Tools Community             启用:
mysql-tools-community-source       MySQL Tools Community - Source    禁用
mysql-tools-preview/x86_64         MySQL Tools Preview               禁用
mysql-tools-preview-source         MySQL Tools Preview - Source      禁用
mysql55-community/x86_64           MySQL 5.5 Community Server        禁用
mysql55-community-source           MySQL 5.5 Community Server - Sour 禁用
mysql56-community/x86_64           MySQL 5.6 Community Server        禁用
mysql56-community-source           MySQL 5.6 Community Server - Sour 禁用
mysql57-community/x86_64           MySQL 5.7 Community Server        启用:
mysql57-community-source           MySQL 5.7 Community Server - Sour 禁用
mysql80-community/x86_64           MySQL 8.0 Community Server        禁用
mysql80-community-source           MySQL 8.0 Community Server - Sour 禁用
[root@huatec01 yum]#
```

图4-5 MySQL仓库版本系列

如果要修改默认的安装版本系列，那么可以使用yum-config-manager指令进行修改。这里选择默认的版本安装即可，安装指令如下：

```
[root@huatec01 yum]# yum install mysql-community-server
```

MySQL安装示意图如图4-6所示。

mysql-community-server指令会下载并安装mysql-community-server、mysql-community-client、mysql-community-common、mysql-community-libs等安装包，需要花费一定的时间。

```
zhusheng — root@huatec01:/etc/yum — ssh root@huatec01 — 80×22
...root@huatec01:/etc/yum — ssh root@huatec01                ~
[root@huatec01 yum]# yum install mysql-community-server
已加载插件： fastestmirror
Loading mirror speeds from cached hostfile
 * base: centos.ustc.edu.cn
 * extras: centos.ustc.edu.cn
 * updates: centos.ustc.edu.cn
正在解决依赖关系
--> 正在检查事务
---> 软件包 mysql-community-server.x86_64.0.5.7.20-1.el7 将被 安装
--> 正在处理依赖关系 mysql-community-common(x86-64) = 5.7.20-1.el7，它被软件包 m
ysql-community-server-5.7.20-1.el7.x86_64 需要
--> 正在处理依赖关系 mysql-community-client(x86-64) >= 5.7.9，它被软件包 mysql-c
ommunity-server-5.7.20-1.el7.x86_64 需要
--> 正在检查事务
---> 软件包 mysql-community-client.x86_64.0.5.7.20-1.el7 将被 安装
--> 正在处理依赖关系 mysql-community-libs(x86-64) >= 5.7.9，它被软件包 mysql-com
munity-client-5.7.20-1.el7.x86_64 需要
---> 软件包 mysql-community-common.x86_64.0.5.7.20-1.el7 将被 安装
--> 正在检查事务
---> 软件包 mariadb-libs.x86_64.1.5.5.35-3.el7 将被 取代
--> 正在处理依赖关系 libmysqlclient.so.18()(64bit)，它被软件包 2:postfix-2.10.1-
6.el7.x86_64 需要
```

图4-6 MySQL安装示意图

然后启动MySQL并查看MySQL服务状态，代码如下：

```
[root@huatec01 yum] #service mysqld start
[root@huatec01 yum] #service mysqld status
```

启动MySQL服务的效果如图4-7所示。

```
[root@huatec01 yum]# service mysqld status
Redirecting to /bin/systemctl status  mysqld.service
mysqld.service - MySQL Server
   Loaded: loaded (/usr/lib/systemd/system/mysqld.service; enabled)
   Active: active (running) since 四 2017-12-28 14:45:46 CST; 16s ago
     Docs: man:mysqld(8)
           http://dev.mysql.com/doc/refman/en/using-systemd.html
  Process: 2113 ExecStart=/usr/sbin/mysqld --daemonize --pid-file=/var/run/mysql
d/mysqld.pid $MYSQLD_OPTS (code=exited, status=0/SUCCESS)
  Process: 2040 ExecStartPre=/usr/bin/mysqld_pre_systemd (code=exited, status=0/
SUCCESS)
 Main PID: 2117 (mysqld)
   CGroup: /system.slice/mysqld.service
           └─2117 /usr/sbin/mysqld --daemonize --pid-file=/var/run/mysqld/mysqld
.pid

12月 28 14:45:46 huatec01 systemd[1]: Started MySQL Server.
[root@huatec01 yum]# █
```

图4-7 启动MySQL服务

为了方便后续使用MySQL数据库，我们将数据库设置为开机启动，代码如下：

```
[root@huatec01 yum] #systemctl enable mysqld
[root@huatec01 yum] #systemctl daemon-reload
```

MySQL 5.7会自动生成默认的数据库密码（5.6系列安装完后需要手动设置），所以第一次登录MySQL需要找到默认的密码进行登录，查看指令如下：

```
[root@huatec01 yum]# grep 'temporary password' /var/log/mysqld.log
```

再使用默认密码登录MySQL shell客户端，然后修改默认的数据库密码，修改MySQL默认的数据库密码的指令如下：

```
mysql> set password for 'root' @ 'localhost' = password('root');
```

修改数据库密码的效果如图4-8所示。

```
mysql> set password for 'root'@'localhost'=password('root');
Query OK, 0 rows affected, 1 warning (0.00 sec)

mysql> 
```

图4-8 修改数据库密码的效果

为了让外部用户连接到MySQL数据库，可以为数据库设置访问权限，执行代码如下：

```
mysql> GRANT ALL PRIVILEGES ON *.* TO 'root' @ '%' IDENTIFIED BY
    'root' WITH GRANT OPTION;
mysql> GRANT ALL PRIVILEGES ON *.* TO 'zhusheng' @ '%' IDENTIFIED BY
    'zhusheng' WITH GRANT OPTION;
```

在上面的代码中，我们分别为root用户和普通用户zhusheng开放了外部访问权限。

4.2.3 配置Hive

为了方便使用Hive，首先为Hive配置全局环境变量，代码如下：

```
[root@huatec01 bin] #vi /etc/profile
...
#hive
export HIVE_HOME = /huatec/apache-hive-2.1.1-bin
export PATH = $PATH:$HIVE_HOME/bin
```

在环境变量配置文件的结尾添加与Hive有关的环境变量，然后执行“source /etc/profile”让配置更新并生效。

上面为Hive安装了元数据库MySQL，但是还没有进行配置。现在进入Hive安装目录下的conf目录，该目录默认存在hive-default.xml.template，它是Hive的配置模板文件，我们将其名称改为hive-site.xml，然后修改其中的内容，代码如下：

```
...
//499行
<property>
```

```
    <name>javax.jdo.option.ConnectionURL</name>
    <value>jdbc:mysql://huatec01:3306/hive?createDatabaseIfNotExist =
        true</value>
    <description>
        JDBC connect string for a JDBC metastore.
        To use SSL to encrypt/authenticate the connection, provide
            database-specific SSL flag in the connection URL.
        For example,jdbc:postgresql://myhost/db?ssl =
            true for postgres database.
    </description>
</property>
...
//931行
<property>
    <name>javax.jdo.option.ConnectionDriverName</name>
    <value>com.mysql.jdbc.Driver</value>
    <description>Driver class name for a JDBC metastore</description>
</property>
...
//956行
<property>
    <name>javax.jdo.option.ConnectionUserName</name>
    <value>root</value>
    <description>Username to use against metastore database</description>
</property>
...
//484行
<property>
    <name>javax.jdo.option.ConnectionPassword</name>
    <value>root</value>
    <description>password to use against metastore database</description>
</property>
...
//685行
<property>
    <name>hive.metastore.schema.verification</name>
    <value>false</value>
    <description>
        Enforce metastore schema version consistency.
        True:Verify that version information stored in is compatible with
            one from Hive jars.
        Also disable automatic schema migration attempt.Users are
            required to
            manually migrate schema after Hive upgrade which
                ensures proper
            metastore schema migration.(Default)
        False:Warn if the version information stored in metastore
            doesn't match
```

```
        with one from in Hive jars.
    </description>
</property>
```

上面的配置文件中修改了五个属性，第一个属性用于指定数据库连接，Hive在第一次启动的时候会默认创建Hive数据库；第二个属性用于指定MySQL的驱动方式为“com.mysql.jdbc.Driver”；第三个属性和第四个属性分别用于指定连接数据库的用户名和密码；第五个属性的值为元数据库表认证，我们将其值修改为“false”。然后将该配置文件中的“${system:java.io.tmpdir}”全部替换为“/huatec/apache-hive-2.1.1-bin/tmp”，并手动创建该路径。

启动Hive前，初始化元数据库，指令如下：

```
[root@huatec01 bin]# ./schematool -initSchema -dbType mysql
```

初始化Hive元数据库的效果如图4-9所示。

```
[root@huatec01 bin]# ./schematool -initSchema -dbType mysql
SLF4J: Class path contains multiple SLF4J bindings.
SLF4J: Found binding in [jar:file:/huatec/apache-hive-2.1.1-bin/lib/log4j-slf4j-impl-2.4.1.jar!/org/s
lf4j/impl/StaticLoggerBinder.class]
SLF4J: Found binding in [jar:file:/huatec/hadoop-2.7.3/share/hadoop/common/lib/slf4j-log4j12-1.7.10.j
ar!/org/slf4j/impl/StaticLoggerBinder.class]
SLF4J: See http://www.slf4j.org/codes.html#multiple_bindings for an explanation.
SLF4J: Actual binding is of type [org.apache.logging.slf4j.Log4jLoggerFactory]
Metastore connection URL:        jdbc:mysql://huatec01:3306/hive?createDatabaseIfNotExist=true
Metastore Connection Driver :    com.mysql.jdbc.Driver
Metastore connection User:       root
Tue Jan 02 11:44:09 CST 2018 WARN: Establishing SSL connection without server's identity verification
 is not recommended. According to MySQL 5.5.45+, 5.6.26+ and 5.7.6+ requirements SSL connection must
be established by default if explicit option isn't set. For compliance with existing applications not
 using SSL the verifyServerCertificate property is set to 'false'. You need either to explicitly disa
ble SSL by setting useSSL=false, or set useSSL=true and provide truststore for server certificate ver
ification.
Starting metastore schema initialization to 2.1.0
Initialization script hive-schema-2.1.0.mysql.sql
Tue Jan 02 11:44:10 CST 2018 WARN: Establishing SSL connection without server's identity verification
 is not recommended. According to MySQL 5.5.45+, 5.6.26+ and 5.7.6+ requirements SSL connection must
be established by default if explicit option isn't set. For compliance with existing applications not
 using SSL the verifyServerCertificate property is set to 'false'. You need either to explicitly disa
ble SSL by setting useSSL=false, or set useSSL=true and provide truststore for server certificate ver
ification.
Initialization script completed
Tue Jan 02 11:44:12 CST 2018 WARN: Establishing SSL connection without server's identity verification
 is not recommended. According to MySQL 5.5.45+, 5.6.26+ and 5.7.6+ requirements SSL connection must
be established by default if explicit option isn't set. For compliance with existing applications not
 using SSL the verifyServerCertificate property is set to 'false'. You need either to explicitly disa
ble SSL by setting useSSL=false, or set useSSL=true and provide truststore for server certificate ver
ification.
schemaTool completed
```

图4-9　初始化Hive元数据库的效果

当控制台输出“schemaTool completed”日志时，表明元数据库初始化成功。启动Hive，如图4-10所示，表明Hive成功进入Shell模式。

```
[root@huatec01 tmp]# hive
SLF4J: Class path contains multiple SLF4J bindings.
SLF4J: Found binding in [jar:file:/huatec/apache-hive-2.1.1-bin/lib/l
SLF4J: Found binding in [jar:file:/huatec/hadoop-2.7.3/share/hadoop/c
SLF4J: See http://www.slf4j.org/codes.html#multiple_bindings for an e
SLF4J: Actual binding is of type [org.apache.logging.slf4j.Log4jLogge

Logging initialized using configuration in jar:file:/huatec/apache-hi
Tue Jan 02 11:52:26 CST 2018 WARN: Establishing SSL connection withou
ablished by default if explicit option isn't set. For compliance with
etting useSSL=false, or set useSSL=true and provide truststore for se
Tue Jan 02 11:52:26 CST 2018 WARN: Establishing SSL connection withou
ablished by default if explicit option isn't set. For compliance with
etting useSSL=false, or set useSSL=true and provide truststore for se
Tue Jan 02 11:52:26 CST 2018 WARN: Establishing SSL connection withou
ablished by default if explicit option isn't set. For compliance with
etting useSSL=false, or set useSSL=true and provide truststore for se
Tue Jan 02 11:52:26 CST 2018 WARN: Establishing SSL connection withou
ablished by default if explicit option isn't set. For compliance with
etting useSSL=false, or set useSSL=true and provide truststore for se
Tue Jan 02 11:52:29 CST 2018 WARN: Establishing SSL connection withou
ablished by default if explicit option isn't set. For compliance with
etting useSSL=false, or set useSSL=true and provide truststore for se
Tue Jan 02 11:52:29 CST 2018 WARN: Establishing SSL connection withou
ablished by default if explicit option isn't set. For compliance with
etting useSSL=false, or set useSSL=true and provide truststore for se
Tue Jan 02 11:52:29 CST 2018 WARN: Establishing SSL connection withou
ablished by default if explicit option isn't set. For compliance with
etting useSSL=false, or set useSSL=true and provide truststore for se
Tue Jan 02 11:52:29 CST 2018 WARN: Establishing SSL connection withou
ablished by default if explicit option isn't set. For compliance with
etting useSSL=false, or set useSSL=true and provide truststore for se
Hive-on-MR is deprecated in Hive 2 and may not be available in the fu
hive>
```

图4-10 Hive shell模式示意图

4.3 Hive基本操作

4.3.1 Hive数据类型

Hive支持两种数据类型，一种叫原子数据类型，即基本数据类型，一种叫复杂数据类型。基本数据类型包括数值型、布尔型和字符串类型，具体如表4-2所示。

表4-2 Hive数据类型

基本数据类型		
类　型	描　述	示　例
TINYINT	1字节（8位），有符号整数	1
SMALLINT	2字节（16位），有符号整数	1
INT	4字节（32位），有符号整数	1
BIGINT	8字节（64位），有符号整数	1
FLOAT	4字节（32位），单精度浮点数	1.0
DOUBLE	8字节（64位），双精度浮点数	1.0
BOOLEAN	True/False	True
STRING	字符串	'xiao'、"xiao"

从表4-2中可以看出，Hive没有定义支持日期的数据类型，在Hive中，日期都是用字符串来表示的，而常用的日期格式转化操作则是通过自定义函数进行定义的。

Hive是使用Java语言开发的，Hive里的基本数据类型和Java中的基本数据类型也是一一对应的，除了string类型。有符号的整数类型TINYINT、SMALLINT、INT和BIGINT分别等价于Java中的byte、short、int和long基本类型，它们分别为1字节、2字节、4字节和8字节有符号整数。Hive的浮点数据类型FLOAT和DOUBLE对应于Java中的float和double类型。而Hive的BOOLEAN类型相当于Java的基本数据类型boolean。对于Hive的STRING类型相当于数据库的varchar类型，该类型是一个可变的字符串，但是它不能声明其中最多能存储多少个字符，理论上它可以存储2 GB的字符数。

Hive支持基本数据类型的转换，低字节的基本数据类型可以转化为高字节的基本数据类型，例如TINYINT、SMALLINT、INT可以转化为FLOAT，而所有的整数类型FLOAT以及STRING类型可以转化为DOUBLE类型，这些转化可以从Java语言的类型转化考虑，因为Hive就是使用Java语言编写的。当然也支持高字节类型转化为低字节类型，这就需要使用Hive的自定义函数CAST了。复杂数据类型包括数组（ARRAY）、映射（MAP）和结构体（STRUCT），具体如表4-3所示。

表4-3 Hive中的复杂数据类型

复杂数据类型		
类 型	描 述	示 例
ARRAY	一组有序字段。字段的类型必须相同	ARRAY(1,2)
MAP	一组无序的键/值对。键的类型必须是原子的，值可以是任何类型，同一个映射的键的类型必须相同，值的类型也必须相同	MAP('a',1,'b',2)
STRUCT	一组命名的字段。字段的类型可以不同	STRUCT('a',1,1,0)

4.3.2 Hive常见函数

微课：
v4-1 Hive
数据类型

Hive语句支持函数式编程，Hive中常见的函数包括数学函数、集合函数和条件函数等。Hive中的数学函数如表4-4所示，Hive中的集合函数如表4-5所示，Hive中的条件函数如表4-6所示。

表4-4 Hive中的数学函数

返回值类型	函数名称	描 述
DOUBLE	round(DOUBLE a)	返回对a四舍五入的BIGINT值
DOUBLE	round(DOUBLE a,INT d)	返回DOUBLE类型的d的保留n位小数的DOUBLE类型的近似值
DOUBLE	bround(DOUBLE a)	银行家舍入法（1~4表示舍去，6~9表示进上，5前位数是偶数表示舍5不进，5前位数是奇数表示舍5入1）
DOUBLE	bround(DOUBLE a,INT d)	银行家舍入法，保留d位小数
BIGINT	floor(DOUBLE a)	向下取整，即取按照数轴上最接近要求值的左边值，如6.10→6 -3.4→-4
BIGINT	ceil(DOUBLE a)、ceiling(DOUBLE a)	求其不小于给定实数的最小整数，如ceil(6) = ceil(6.1)= ceil(6.9) = 6

续表

返回值类型	函数名称	描 述
DOUBLE	rand()、rand(INT seed)	每行返回一个DOUBLE类型的随机数，seed是随机因子
DOUBLE	exp(DOUBLE a)、exp(DECIMAL a)	返回e的a次幂，a可为小数
DOUBLE	ln(DOUBLE a)、ln(DECIMAL a)	以自然数为底的对数，a可为小数
DOUBLE	log10(DOUBLE a)、log10(DECIMAL a)	以10为底的对数，a可为小数
DOUBLE	log2(DOUBLE a)、log2(DECIMAL a)	以2为底的对数，a可为小数
DOUBLE	log(DOUBLE base,DOUBLE a)、log(DECIMAL base,DECIMAL a)	以base为底的对数，base与a都是DOUBLE类型
DOUBLE	pow(DOUBLE a,DOUBLE p)、power(DOUBLE a,DOUBLE p)	计算a的p次幂
DOUBLE	sqrt(DOUBLE a)、sqrt(DECIMAL a)	计算a的平方根
STRING	bin(BIGINT a)	计算二进制a的STRING类型、a为BIGINT类型
STRING	hex(BIGINT a)、hex(STRING a)、hex(BINARY a)	计算十六进制a的STRING类型，如果a为STRING类型，就转换成字符相对应的十六进制
STRING	conv(BIGINT num,INT from_base,INT to_base)、conv(STRING num,INT from_base,INT to_base)	将GIGINT/STRING类型的num从from_base进制转换成to_base进制
DOUBLE	abs(DOUBLE a)	计算a的绝对值
INT或DOUBLE	pmod(INT a,INT b)、pmod(DOUBLE a,DOUBLE b)	a对b取模
DOUBLE	sin(DOUBLE a)、sin(DECIMAL a)	求a的正弦值
DOUBLE	asin(DOUBLE a)、asin(DECIMAL a)	求a的反正弦值
DOUBLE	cos(DOUBLE a)、cos(DECIMAL a)	求余弦值
DOUBLE	acos(DOUBLE a)、acos(DECIMAL a)	求反余弦值
DOUBLE	tan(DOUBLE a)、tan(DECIMAL a)	求正切值
DOUBLE	atan(DOUBLE a)、atan(DECIMAL a)	求反正切值
DOUBLE	degrees(DOUBLE a)、degrees(DECIMAL a)	将弧度值转换成角度值
DOUBLE	radians(DOUBLE a)、radians(DOUBLE a)	将角度值转换成弧度值
INT或DOUBLE	positive(INT a)、positive(DOUBLE a)	返回a
INT或DOUBLE	negative(INT a)、negative(DOUBLE a)	返回a的相反数
DOUBLE或INT	sign(DOUBLE a)、sign(DECIMAL a)	如果a是正数，则返回1.0，如果是负数，则返回-1.0，否则返回0.0
DOUBLE	e()	数学常数e
DOUBLE	pi()	数学常数π
BIGINT	factorial(INT a)	求a的阶乘
DOUBLE	cbrt(DOUBLE a)	求a的立方根
INT BIGINT	shiftleft(TINYINT|SMALLINT|INT a,INT b)、shiftleft(BIGINT a,INT b)	按位左移

续表

<table>
<tr><th>返回值类型</th><th>函数名称</th><th>描 述</th></tr>
<tr><td>INT
BIGINT</td><td>shiftright(TINYINT|SMALLINT|INT a,INT b、)
shiftright(BIGINT a,INT b)</td><td>按位右移</td></tr>
<tr><td>INT
BIGINT</td><td>shiftrightunsigned(TINYINT|SMALLINT|INT a,INT b),
shiftrightunsigned(BIGINT a,INT b)</td><td>无符号按位右移（<<<）</td></tr>
<tr><td>T</td><td>greatest(T v1,T v2,...)</td><td>求最大值</td></tr>
<tr><td>T</td><td>least(T v1,T v2,...)</td><td>求最小值</td></tr>
</table>

表4-5 Hive中的集合函数

返回值类型	函数名称	描述
int	size(Map<K.V>)	求map的长度
int	size(Array<T>)	求数组的长度
array<K>	map_keys(Map<K.V>)	返回map中的所有key
array<V>	map_values(Map<K.V>)	返回map中的所有value
boolean	array_contains(Array<T>,value)	如果该数组Array<T>包含value，则返回True，否则返回False
array	sort_array(Array<T>)	按自然顺序对数组进行排序并返回

表4-6 Hive中的条件函数

返回值类型	函数名称	描述
T	if(boolean testCondition,T valueTrue,T valueFalseOrNull)	如果testCondition为True，则返回valueTrue,否则返回valueFalseOrNull（valueTrue、valueFalseOrNull为泛型）
T	nvl(T value,T default_value)	如果value值为NULL，则返回default_value，否则返回value
T	COALESCE(T v1,T v2,...)	返回第一非NULL的值，如果全部为NULL，就返回NULL，如COALESCE (NULL,44,55)=44/strong>
T	CASE a WHEN b THEN c [WHEN d THEN e]* [ELSE f] END	如果a=b，就返回c，如果a=d，就返回e，否则返回f，如CASE 4 WHEN 5 THEN 5 WHEN 4 THEN 4 ELSE 3 END将返回4
T	CASE WHEN a THEN b [WHEN c THEN d]* [ELSE e] END	如果a=True，就返回b,如果c=True，就返回d，否则返回e，如CASE WHEN 5 >0 THEN 5 WHEN 4>0 THEN 4 ELSE 0 END将返回5；CASE WHEN 5<0 THEN 5 WHEN 4<0 THEN 4 ELSE 0 END将返回0
boolean	isnull(a)	如果a为NULL，就返回True，否则返回False
boolean	isnotnull(a)	如果a为非NULL，就返回True，否则返回False

4.3.3 Hive表操作

Hive进行数据分析时需要确保我们的数据是使用Hive进行管理和维护的，Hive管理数据是采用表的方式进行的。前面已经提到Hive的表分为四种类型，即内部表、外部表、分区表和桶表。

执行所有的表操作需要在Hive Shell模式下进行，所以需要先成功进入Hive Shell模式，然后调用Hive提供的Shell语法执行表操作。

Hive Shell除了四种表的创建操作比较特殊外，其他操作方法与MySQL的操作方法类似，因此，这里将结合相关的示例说明这四种表是如何创建并加载数据的。

1.创建数据库

创建数据库的代码如下：

```
hive> create database db_name;   //创建名为db_name的数据库
hive> show databases;            //显示当前数据库
hive> use db_name;  //进入名为db_name的数据库（注意：若没有此条语句，则hive表存
                    放在hive默认的数据库default中）
```

2.创建表

在Hive中，数据表的类型分为外部表和内部表，以下为创建数据表的Hive语句：

```
CREATE [EXTERNAL] TABLE [IF NOT EXISTS] table_name
[(col_name data_type [COMMENT col_comment], ...)]
[COMMENT table_comment]
[PARTITIONED BY (col_name data_type [COMMENT col_comment], ...)]
[CLUSTERED BY (col_name, col_name, ...)
[SORTED BY (col_name [ASC|DESC], ...)] INTO num_buckets BUCKETS]
[ROW FORMAT row_format]
[STORED AS file_format]
[LOCATION hdfs_path]
ROW FORMAT DELIMITED
[FIELDS TERMINATED BY char]
[COLLECTION ITEMS TERMINATED BY char]
[MAP KEYS TERMINATED BY char]
[LINES TERMINATED BY char]
| SERDE serde_name [WITH SERDEPROPERTIES (property_name=property_value,
property_name=property_value, ...)]
STORED AS
SEQUENCE FILE
| TEXTFILE
| RCFILE
| INPUTFORMAT input_format_classname OUTPUTFORMAT output_format_
classname
```

其中：CREATE TABLE表示创建一个指定名字的表，如果相同名字的表已经存在，则抛出异常；用户可以用IF NOT EXISTS选项来忽略这个异常；EXTERNAL关键字可以让用户创建一个外部表，在创建表的同时指定一个指向实际数据的路径（LOCATION）；LIKE允许用户复制现有的表结构，但是不复制数据；COMMENT可以为表与字段增加描述。

用户在创建表的时候可以自定义SerDe或者使用自带的SerDe。如果没有指定ROW FORMAT或者ROW FORMAT DELIMITED，则将会使用自带的SerDe。在创建表的时候，用户还需要为表指定列，用户在指定表的列的同时也会指定自定义的SerDe，Hive通过SerDe确定表的具体的列的数据。如果文件数据是纯文本，则可以使用STORED AS TEXTFILE。如果数据需要压缩，则使用STORED AS SEQUENCE。

3. 内部表

微课：
v4-2 创建内部表

内部表是很常用的一种Hive表类型，首先我们创建一个内部表m_data，创建的表语句如下所示：

```
hive> create table m_data(country string,gold int,silver
    int,copper int) row format delimited
fields terminated by '';
```

运行上面的创建表语句，效果如图4-11所示。

```
hive> create table m_data(country string,gold int,silver int,copper int)
    > row format delimited
    > fields terminated by'\t';
OK
Time taken: 1.504 seconds
```

图4-11 创建内部表语句的效果

在创建完内部表后，执行“show tables;”命令会显示所有的表，可以看到我们新建的表“m_data”位于其中。新建的Hive表是一个没有任何数据集的空表，我们将本地的数据加载到该表中，数据加载指令如图4-12所示。然后使用与MySQL完全一样的select查询语句来查询数据。

```
hive> show tables;
OK
t_order
Time taken: 0.131 seconds, Fetched: 1 row(s)
```

图4-12 数据加载指令

对2022年北京冬奥会获奖数据进行分析，截至2022年2月24日，北京冬奥会各国家/地区奖牌获得情况数据如表4-7所示。

表4-7 北京冬奥会各国家/地区奖牌获得情况数据

国家/地区	金牌(Gold)	银牌(Silver)	铜牌(Copper)
Norway	16	8	13
Germany	12	10	5
China	9	4	2
America	8	10	7
Sweden	8	5	5
Holland	8	5	4
Austrians	7	7	4
Switzerland	7	2	5

为了让Hive更好地分析以上数据，我们将以上数据存放到Linux操作系统中，命令如下所示：

```
Norway 16 8 13
Germany 12 10 5
China 9 4 2
America 8 10 7
Sweden 8 5 5
Holland 8 5 4
```

```
Austrians 7 7 4
Switzerland 7 2 5
```

以上命令的运行效果如图4-13所示。

```
[root@localhost ~]# vi m_data.txt
Norway 16 8 13
Germany 12 10 5
China 9 4 2
America 8 10 7
Sweden 8 5 5
Holland 8 5 4
Austrians 7 7 4
Switzerland 7 2 5
```

图4-13 运行效果

微课：
v4-3 加载数据到内部表

内部表、m_data.txt文件创建完成及数据录入之后，需要加载数据到m_data内部表中，命令如下：

```
hive> load data local inpath '/root/m_data.txt' into table m_data;
```

将数据加载到我们创建的Hive内部表之后，可以通过select查询语句查询数据，执行效果如图4-14所示。

微课：
v4-4 覆盖表数据

```
hive> load data local inpath'/root/m_data.txt'into table m_data;
Loading data to table db_name.m_data
Table db_name.m_data stats: [numFiles=1, totalSize=119]
OK
Time taken: 0.98 seconds
hive> select *from m_data;
OK
Norway  16      8       13
Germany 12      10      5
China   9       4       2
America 8       10      7
Sweden  8       5       5
Holland 8       5       4
Austrians       7       7       4
Switzerland     7       2       5
Time taken: 0.382 seconds, Fetched: 8 row(s)
```

图4-14 内部表数据的加载

经过前面的讲解，我们知道Hive的底层是依赖HDFS的，也就是说，Hive的数据实际上还是存储在HDFS上的，而且这个存储目录也可以通过hive-site.xml进行设置。在浏览器中访问HDFS页面进行验证的效果如图4-15所示。

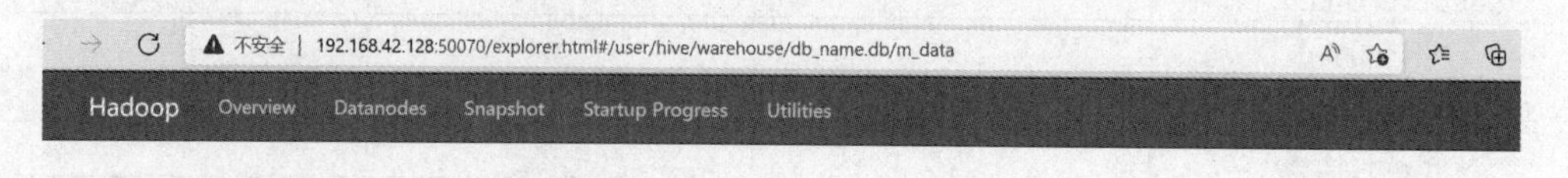

Browse Directory

/user/hive/warehouse/db_name.db/m_data

Permission	Owner	Group	Size	Last Modified	Replication	Block Size	Name
-rwxr-xr-x	root	supergroup	119 B	2022/3/26 20:34:14	2	128 MB	m_data.txt
-rwxr-xr-x	root	supergroup	119 B	2022/3/26 20:37:09	2	128 MB	m_data_copy_1.txt

图4-15 在浏览器中访问HDFS页面进行验证的效果

从图4-15可以看出，在“/user/hive/warehouse/db_name.db/m_data”目录下有一个文件“m_data.txt”，这再次表明，Hive表在HDFS上对应的其实是一个目录，目录下的所有文件都是表的数据，这种存储结构和MySQL是完全不同的，我们需要加以理解。

4. 外部表

EXTERNAL关键字可以让用户创建一个外部表，在创建表的同时指定一条指向实际数据的路径（LOCATION），Hive创建内部表时，会将数据移动到数据仓库指向的路径。若创建外部表时仅记录数据所在的路径，不对数据的位置进行任何改变，则在删除表的时候，内部表的元数据和数据会被一起删除，而外部表只删除元数据，不删除数据。通过内部表的创建和使用，可以知道内部表的数据其实就是存储在HDFS上的，而表中的数据是我们通过本地导入的。如果现在数据本身就存储在HDFS上，那么是否可以使用Hive来管理呢？难道要先导出到本地，然后导入Hive中？其实不用这么麻烦，Hive在设计的时候已经为我们考虑到了这一点，只需要使用Hive的外部表就可以轻松实现这种业务需求。但是，外部表管理数据必须在创建表的同时就指定外部数据的位置。理论上，只要数据表格式和数据格式都是标准化的，而且一一对应，不管数据存储在HDFS的什么位置，都可以使用Hive进行管理。

现在使用外部表的方式将其纳入Hive的管理中去，创建外部表并加载数据的整个执行语句如下所示：

```
hive> create external table order_log
    > (order_id int,sn string,member_id int,
    > status int,payment_id int,logi_id int,
    > total_amount double,address_id int,
    > create_time string,modify_time string)
    > row format delimited fields terminated by '\t'
> location '/mobileshop';
OK
```

需要注意的是，location后面跟的是HDFS上的一个目录，而不是具体的某个文件。比如order-20211021.log文件在HDFS上的绝对路径为hdfs://ns1/mobileshop/order-20211021.log，但是指定location的时候只需要告诉order-20211021.log文件所在的目录即可，也就是hdfs://ns1/mobileshop。我们也可以使用相对路径表示“/mobileshop”。

这说明了一个问题，如果/mobileshop路径下有很多与order-20211021.log文件结构相同的日志文件，则上述指令将会遍历该目录下的所有文件，然后都关联到Hive外部表中去。

创建外部表并加载数据的执行效果如图4-16所示。

```
hive> create external table order_log
    > (order_id int, sn string, member_id int,
    > status int, payment_id int, logi_id int,
    > total_amount double, address_id int,
    > create_time string, modify_time string)
    > row format delimited fields terminated by '\t'
    > location '/mobileshop ';
OK
Time taken: 0.035 seconds
hive> load data local inpath '/home/hw18883508548/c_323/order_log.txt'into table order_log;
Loading data to table liaoling_.order_log
OK
Time taken: 0.249 seconds
```

图4-16 创建外部表并加载数据的执行效果

微课：
v4-5 创建外部表

从图4-16中可以看出，执行操作是成功的，当执行select * from order_log;查询语句的时候，能够查询到相应的数据，如图4-17所示。

```
hive> select *from order_log;
OK
1       zhangsan        1       9       202232301       32301   56.0
2       lisi    2       8       202232302       32302   59.0    2302
1       zhangsan        1       9       202232301       32301   56.0
2       lisi    2       8       202232302       32302   59.0    2302
Time taken: 0.22 seconds, Fetched: 4 row(s)
```

图4-17 外部表order_log中的数据

5. 分区表

分区表是为了方便快速查询数据而设计的表。可以按照学生信息表的入学时间等对数据进行分类管理，比如一个时间是一个目录层级，省下的每个市是一个子目录层级，该信息表的数据都存储在该目录下。也可以将学生信息表的数据继续按照年月来进行细分管理。当我们需要查询例如“不同年份的数据”时，就可以快速定位到该位置获得返回的数据。分区表可以结合内部表和外部表一起使用，所以分区表的操作也分为两种。首先说明内部表是如何使用分区操作的，思路是在创建表的时候指定分区；然后将数据导入表中。创建分区表的语句如下：

```
hive> create table table_name
    > (id int comment '学号',
    > name string comment '姓名')
    > partitioned by(year int)
    > row format delimited
    > fields terminated by '';
```

执行上述指令，创建内部分区表的执行效果如图4-18所示。

```
Time taken: 1.172 seconds
hive> create table table_name
    > (id int comment '学号',
    > name string comment '姓名')
    > partitioned by(year int)
    > row format delimited
    > fields terminated by '  ';
OK
Time taken: 0.341 seconds
```

图4-18 创建内部分区表的执行效果

内部分区表的创建和内部表的创建的区别在于需要指定“partitioned by”，通俗来说就是分区字段。需要注意的是，分区字段名称不能是表的字段名称。例如，在上面创建的表语句中，我们重新取了一个分区字段名称“year”来代表不同年份入学的时间。

接下来为表加载数据，加载指令如下：

```
hive> load data local inpath
    > '/home/hw18883508548/year_2020.txt'
    > overwrite into table table_name partition(year=2020);
Loading data to table d_323.table_name partition (year=2020)
OK
Time taken: 0.467 seconds
hive> load data local inpath
    > '/home/hw18883508548/year_2021.txt'
    > overwrite into table table_name partition(year=2021);
Loading data to table d_323.table_name partition (year=2021)
OK
Time taken: 0.349 seconds
```

执行上述导入数据的指令，在本地目录“/home/hw18883508548”下有几个文件“year_2020”、“year_2021”，它们分别表示2020年和2021年入学的学生信息数据。我们要分别指定分区名称，然后导入，这样导入后的数据才能以分区名称形成自己的目录层级。图4-19为加载2020年和2021年的学生表数据到分区表的执行结果。

```
hive> load data local inpath
    > '/home/hw18883508548/year_2020.txt'
    > overwrite into table table_name partition(year=2020);
Loading data to table d_323.table_name partition (year=2020)
OK
Time taken: 0.467 seconds
hive> load data local inpath
    > '/home/hw18883508548/year_2021.txt'
    > overwrite into table table_name partition(year=2021);
Loading data to table d_323.table_name partition (year=2021)
OK
Time taken: 0.349 seconds
```

图4-19 加载2020年和2021年的学生表数据到分区表的执行结果

从图4-19中可以看出，执行操作是成功的，当执行“select * from table_name;”查询语句的时候，能够查询到相应的不同时间入学的学生信息数据，如图4-20所示。

```
hive> select *from table_name;
OK
1       zhangsan        2020
2       lisi    2020
3       tony    2020
4       jack    2020
5       kangkang        2020
1       wangwu  2021
2       lily    2021
3       wanger  2021
4       lucy    2021
5       xiaowang        2021
Time taken: 1.956 seconds, Fetched: 10 row(s)
```

图4-20 查询分区表中的学生信息数据

微课：
v4-6 分区表

由HDFS分区层级关系可知，一个分区代表一个文件夹，该文件夹下为分区的数据文件。虽然分区字段在HDFS上表示的是一个层级目录，但它还是表的字段，所以完整的分区表字段是包含分区字段的，HDFS上的分区文件目录结构如图4-21所示。

```
[hw18883508548@ddhvd ~]$ hadoop fs -ls hdfs://masters/user/hive/warehouse/d_323.db/table_name
22/03/24 09:56:53 WARN util.NativeCodeLoader: Unable to load native-hadoop library for your platform... using builtin-java classes where
plicable
Found 2 items
drwxr-xr-x   - hw18883508548 supergroup          0 2022-03-23 11:24 hdfs://masters/user/hive/warehouse/d_323.db/table_name/year=2020
drwxr-xr-x   - hw18883508548 supergroup          0 2022-03-23 11:25 hdfs://masters/user/hive/warehouse/d_323.db/table_name/year=2021
```

图4-21 分区文件目录结构

可以使用关键字where来指定分区字段直接作为查询条件，例如“select * from table_name where year=2021;”，执行效果如图4-22所示。

```
hive> select * from table_name where year=2021;
OK
1       wangwu  2021
2       lily    2021
3       wanger  2021
4       lucy    2021
5       xiaowang        2021
Time taken: 3.11 seconds, Fetched: 5 row(s)
```

图4-22 使用关键字where来指定分区字段直接作为查询条件

分区表还可以结合外部表一起使用，它的用法和内部表类似。如果要使用外部分区表，则存储在HDFS上的数据层级结构和数据结构都需要符合分区表的特点。主要思路是先创建一个外部表并指定分区字段，然后为外部分区表添加分区，添加的时候指定数据在HDFS上的存储路径。

创建外部分区表的指令如下：

```
hive> create external table ext_table_name
    > (name string,num string)
    > partitioned by (year int)
    >  row format delimited
```

```
    > fields terminated by ''
> location '/ext_table_address';
```

创建外部分区表的执行效果如图4-23所示。

```
hive> create external table ext_table_name
    > (name string,num string)
    > partitioned by (year int)
    >  row format delimited
    > fields terminated by ' '
    > location '/ext_table_address';
OK
Time taken: 0.372 seconds
```

图4-23 创建外部分区表的执行效果

微课：v4-7 分区目录

需要注意的是，我们在创建外部分区表的时候必须指定location属性，因为外部分区表也是外部表的一种，因此同样需要遵循外部表的创建规则。这里我们指定的HDFS路径为/ext_table_address，HDFS的文件结构如图4-24所示。

```
[hw18883508548@ddhvd ~]$ hadoop fs -ls hdfs://masters/
22/03/24 08:18:51 WARN util.NativeCodeLoader: Unable to load native-hadoop library for your platform..
plicable
Found 26 items
-rw-r--r--   3 hw13628492892 supergroup          0 2021-03-11 15:06 hdfs://masters/1.txt
drwxr-xr-x   - hw13554327709 supergroup          0 2021-09-09 15:33 hdfs://masters/10086
drwxr-xr-x   - hw13986085482 supergroup          0 2021-09-09 16:48 hdfs://masters/11
drwxr-xr-x   - hw13986085482 supergroup          0 2021-09-09 15:34 hdfs://masters/11111111111
drwxr-xr-x   - hw13594706485 supergroup          0 2022-01-04 21:53 hdfs://masters/data
drwxr-xr-x   - hw13594706485 supergroup          0 2022-01-04 21:38 hdfs://masters/data1
drwxr-xr-x   - root          supergroup          0 2022-01-05 17:44 hdfs://masters/ddhome
drwxr-xr-x   - hw13541727214 supergroup          0 2021-11-04 17:17 hdfs://masters/employee
drwxr-xr-x   - hw18883508548 supergroup          0 2022-03-23 12:43 hdfs://masters/ext_table_address
drwxr-xr-x   - root          supergroup          0 2022-03-05 18:39 hdfs://masters/hbase
drwxr-xr-x   - hw18209334539 supergroup          0 2021-10-29 15:25 hdfs://masters/hive_db
drwxr-xr-x   - hw18883508548 supergroup          0 2022-01-05 15:51 hdfs://masters/home
```

图4-24 HDFS的文件结构

6. 桶表

前面已经介绍了内部表、外部表，以及使用分区表对数据进行细分管理。对于每一个表或者分区表，Hive还可以进行更为细颗粒的数据划分和管理，也就是桶。Hive也是基于对某一列进行桶的组织，Hive采用对列值哈希，然后除以桶的个数求余的方式决定该条记录存放在哪个桶当中。将表（或者分区）组织成桶有以下两个优点。

（1）获得更高的查询处理效率。桶为表加上了额外的结构，Hive在处理有些查询的时候能利用这个结构。具体而言，要在两个（包含连接列的）相同列上连接划分了桶的表，可以使用Map端连接（Map-Side Join）实现，比如JOIN操作。对于JOIN操作，两个表有一个相同的列，如果对这两个表都进行桶操作，那么将保存相同列值的桶进行JOIN操作就可，这大大减少了JOIN的数据量。

（2）使取样更高效。在处理大规模数据集的时候，在开发和修改查询的阶段，如果能在数据集的一小部分数据上试运行查询，则会带来很大方便。

使用桶表之前，我们需要进行相关设置，否则输出只有一个文件。设置指令如下：

```
set hive.enforce.bucketing = true
```

要向分桶表中填充数据，需要将hive.enforce.bucketing属性设置为True。这样，Hive就能知道用表定义中声明的数量来创建桶，然后使用INSERT命令即可。创建桶表的语句如下：

```
create table if not exists bk_table_name
(id int comment '学号',
name string comment '姓名')
clustered by(id) into 4 buckets
row format delimited
fields terminated by '';
```

创建桶表的语句的执行效果如图4-25所示。

```
hive> create table if not exists bk_table_name
    > (id int comment '学号',
    > name string comment '姓名')
    > clustered by(id) into 4 buckets
    > row format delimited
    > fields terminated by ' ' ;
OK
Time taken: 0.565 seconds
```

图4-25 创建桶表的语句的执行效果

前面以表“order_log”为例介绍了外部表的创建，并基于该外部表的数据创建桶表。需要注意的是，我们需要对表“order_log”进行桶表管理，那么桶表的表结构必须与表“order_log”的表结构一致。但是，我们新建的桶表是空的，因此需要将“table_name”表的数据导入“bk_table_name”中，指令如下所示：

```
set hive.enforce.bucketing = true;
insert into table bk_table_name select * from s_table_name;
```

为桶表添加数据指令的执行效果如图4-26所示。

```
hive> insert into table bk_table_name select * from s_table_name;
WARNING: Hive-on-MR is deprecated in Hive 2 and may not be available in the future versions. C
.e. spark, tez) or using Hive 1.X releases.
Query ID = hw18883508548_20220324102717_419366a6-aaed-43b7-ad3c-44a62bd82c41
Total jobs = 1
Launching Job 1 out of 1
Number of reduce tasks determined at compile time: 4
In order to change the average load for a reducer (in bytes):
  set hive.exec.reducers.bytes.per.reducer=<number>
In order to limit the maximum number of reducers:
  set hive.exec.reducers.max=<number>
In order to set a constant number of reducers:
  set mapreduce.job.reduces=<number>
Starting Job = job_1646476607176_0161, Tracking URL = http://ddcvb:28088/proxy/application_164
Kill Command = /ddhome/bin/hadoop/bin/hadoop job  -kill job_1646476607176_0161
Hadoop job information for Stage-1: number of mappers: 1; number of reducers: 4
2022-03-24 10:27:45,613 Stage-1 map = 0%,  reduce = 0%
2022-03-24 10:27:59,689 Stage-1 map = 100%,  reduce = 0%, Cumulative CPU 2.37 sec
2022-03-24 10:28:13,294 Stage-1 map = 100%,  reduce = 50%, Cumulative CPU 7.12 sec
2022-03-24 10:28:15,385 Stage-1 map = 100%,  reduce = 75%, Cumulative CPU 9.74 sec
2022-03-24 10:28:16,432 Stage-1 map = 100%,  reduce = 100%, Cumulative CPU 12.23 sec
MapReduce Total cumulative CPU time: 12 seconds 230 msec
```

图4-26 为桶表添加数据指令的执行效果

从图4-26中可以看出，Hive将数据导入操作转换为MapReduce任务，一共有4个“reducers”，可以在HDFS上查看桶表的结构，如图4-27所示。

```
[hw18883508548@ddhwd ~]$ hadoop fs -ls hdfs://masters/user/hive/warehouse/d_323.db/bk_table_name
22/03/24 10:29:52 WARN util.NativeCodeLoader: Unable to load native-hadoop library for your platform... using builtin-java classes where
plicable
Found 4 items
-rwxr-xr-x   3 hw18883508548 supergroup         10 2022-03-24 10:28 hdfs://masters/user/hive/warehouse/d_323.db/bk_table_name/000000_0
-rwxr-xr-x   3 hw18883508548 supergroup         11 2022-03-24 10:28 hdfs://masters/user/hive/warehouse/d_323.db/bk_table_name/000001_0
-rwxr-xr-x   3 hw18883508548 supergroup          7 2022-03-24 10:28 hdfs://masters/user/hive/warehouse/d_323.db/bk_table_name/000002_0
-rwxr-xr-x   3 hw18883508548 supergroup          9 2022-03-24 10:28 hdfs://masters/user/hive/warehouse/d_323.db/bk_table_name/000003_0
```

图4-27 在HDFS上查看桶表结构

微课：
v4-8 桶表

从图4-27中可以看到桶表有4个文件，前面reducers的个数和这里的桶表文件个数都与我们创建桶表时设置的buckets数量有关。

桶表也是Hive表的一种，同样可以使用select等语句进行数据查询，例如select * from bk_table_name，除这些基本的查询操作外，还可以将桶表作为查询条件。比如对桶表数据进行取样，相关指令格式如下：

```
select * from bk_table_name TABLESAMPLE(BUCKET x OUT OF y);
```

其中：x和y是参数，y尽可能是桶表数的倍数或者因子，而且y必须大于x，当执行相关指令时，Hive会根据y来决定抽样的比例，x表示从哪个桶开始进行抽样。

例如clustered by(id) into 16 buckets，表示table总共分了16桶，当y=8时，抽取(16/8=)2个bucket的数据。

以我们创建的桶表bk_table_name为例，桶表共有4个，那么，如果抽取相同的数据呢？常见的抽样操作如下所示。

（1）从bk_order_log分桶表中抽取1桶数据：

假设x=2，y=4，则语句如下：

```
select * from bk_table_name TABLESAMPLE(BUCKET 2 OUT OF 4);
```

执行上述抽样语句，效果如图4-28所示。

```
hive> select * from bk_table_name TABLESAMPLE(BUCKET 2 OUT OF 4);
OK
1       zhangsan
Time taken: 0.29 seconds, Fetched: 1 row(s)
```

图4-28 桶表抽样语句的执行效果

从图4-28可以看到，抽样查询和普通查询的区别就在于我们的数据是样本数据，当数据的基数非常大的时候，可以基于抽样的数据进行抽样调查。

下面再结合bk_table_name表，继续给出抽样2桶、4桶、半桶数据的抽样语句，也可以自己进行查看。当然，如果桶数非常多，这样就可以有更多的抽样选择。

（2）从bk_order_log分桶表中抽取2桶数据：

假设x=2，y=2，则语句如下：

```
select * from bk_order_log TABLESAMPLE(BUCKET 2 OUT OF 2);
```

（3）从bk_order_log分桶表中抽取4桶数据：

假设x=1，y=1，则语句如下：

```
select * from bk_order_log TABLESAMPLE(BUCKET 1 OUT OF 1);
```

（4）从bk_order_log分桶表中抽取半桶数据：

假设x=1，y=8，则语句如下：

```
select * from bk_order_log TABLESAMPLE(BUCKET 1 OUT OF 8);
```

除上述抽样方式外，Hive表还可以通过“by rand() limit x”进行指定条数随机取样，通过“TABLESAMPLE (n PERCENT)”进行块取样，通过“TABLESAMPLE (nM)”指定取样数据大小（单位为MB）等。

桶表除具备以上各种强大的取样功能外，还可以结合分区表进行混合细分管理，由于篇幅有限，此处不一一进行讲解，更多功能请自行查阅相关书籍进行学习。

7. 删除表

在Hive中，删除内部表的同时会删除表的元数据和数据。删除外部表的同时只删除元数据而保留数据。数据库d_323中已包括内部表、外部表、分区表和分桶表，如图4-29所示。

```
hive> use d_323;
OK
Time taken: 0.018 seconds
hive> show tables;
OK
bk_table_name
ext_table_name
par_table_name
s_table_name
table_name
Time taken: 0.051 seconds, Fetched: 5 row(s)
```

图4-29 数据库d_323中已有的表

在Hive中，可以通过ALTER命令删除表，例如，删除表ext_table_name的命令如下：

```
drop table ect_table_name;
```

删除表之后再显示数据库中的表，发现此表已经不存在，此过程如图4-30所示。

```
hive> drop table ext_table_name;
OK
Time taken: 0.122 seconds
hive> show tables;
OK
bk_table_name
par_table_name
s_table_name
table_name
Time taken: 0.057 seconds, Fetched: 4 row(s)
```

图4-30 删除表后再显示数据库d_323中的表

8. 修改表结构

（1）重命名表。

重命名的命令格式如下：

```
ALTER TABLE table_name RENAME TO new_table_name
```

将雇员表employee_external0408修改为employee_ext0408的命令如下：

```
hive> use empdb;
hive> alter table employee_external0408 rename to employee_ext0408;
```

（2）添加和更新列。

添加和更新列的命令格式如下：

```
ALTER TABLE table_name ADD|REPLACE COLUMNS (col_name data_type)
```

其中：ADD表示添加列，REPLACE表示更新列。在雇员表employee_ext0408中增加一列empid，数据类型为string的命令如下：

```
hive> alter table employee_ext0408 add columns(empid string);
```

使用desc命令查看表employee_ext0408的结构，如图4-31所示。

```
hive> desc employee_ext0408;
OK
name                    string
work_place              array<string>
sex_age                 struct<sex:string,age:int>
skill_score             map<string,int>
depart_tile             map<string,array<string>>
empid                   string
Time taken: 0.087 seconds, Fetched: 6 row(s)
```

图4-31 使用desc命令查看表employee_ext0408的结构

在雇员表employee_ext0408中删除新增的列empid的命令如下所示：

```
hive> alter table employee_ext0408 replace columns(
    > name string,
    > work_place ARRAY<string>,
    > sex_age STRUCT<sex:string,age:int>,
    > skills_score MAP<string,int>,
    > depart_title MAP<string,ARRAY<string>>);
```

4.4 Hive高级操作

前面我们详细介绍了Hive的基本操作，所涉及的实际数据类型是基本数据类型。在实际生产环境的数据处理应用中，针对海量的数据，存在大量复杂的数据类型，比如数组、映射和结构体等。在普通的学生信息表中，学生基本信息包括很多字段，比如学号、姓名、身份证号、成绩、课程等。那么，如果想对学生信息表中的结构化数

据的成绩或其他字段进行分析，同样可以采用Hive的复杂数据结构进行相关计算。但对于Hive中的计算问题，Hive需要调用后台的MR进行计算，需要花费的时间较长。

4.4.1 排序

在大量的数据统计中，排序的使用频率较高。Hive中可分为三种不同的排序方式，分别是order by、sort by和distribute by。通常情况下，distribute by和sort by联合使用。例如，关于员工表中的三列数据id、dep_id和sla分别代表id号、部门id号和工资三个字段，实验数据如下：

```
1001 101 8053
1001 102 9852
1001 103 7569
1002 101 9852
1002 102 10255
1002 103 6951
1003 101 6851
1003 102 9654
```

首先将数据存放在Linux本地的/home/hw18883508548/dep_sla.txt目录中，进入Hive中创建表dep_sla，并将本地数据dep_sla.txt加载到dep_sla表中，具体命令如下：

```
hive> use d_323;
OK
Time taken: 0.142 seconds
hive> create table dep_sla(id string,
    > dep_id string,sla int)
    > row format delimited
    > fields terminated by '';
OK
Time taken: 0.498 seconds
hive> load data local inpath '/home/hw18883508548/dep_sla.txt' into
table dep_sla;
Loading data to table d_323.dep_sla
OK
```

创建dep_sla表并加载数据到表中的执行效果如图4-32所示。

```
hive> use d_323;
OK
Time taken: 0.142 seconds
hive> create table dep_sla(id string,
    > dep_id string,sla int)
    > row format delimited
    > fields terminated by' ';
OK
Time taken: 0.498 seconds
hive> load data local inpath'/home/hw18883508548/dep_sla.txt'into table dep_sla;
Loading data to table d_323.dep_sla
OK
Time taken: 0.516 seconds
hive>
```

图4-32 创建dep_sla表并加载数据到表中的执行效果

对于该员工工资表中的sla字段，我们可以根据sla进行升序（asc）和降序（desc）排序，可以用order by关键字来实现，我们把它称为全局排序。HQL命令如下：

```
hive> select * from dep_sla order by sla desc;
```

此排序方式为降序（desc）排序，因为此命令涉及Hive中的排序计算，所以需要调用MR，需要的时间较长，执行结果如图4-33所示。

```
hive> select * from dep_sla order by sla desc;
WARNING: Hive-on-MR is deprecated in Hive 2 and may
.e. spark, tez) or using Hive 1.X releases.
Query ID = hw18883508548_20220325100955_d59136b1-9f
Total jobs = 1
Launching Job 1 out of 1
Number of reduce tasks determined at compile time:
In order to change the average load for a reducer (
  set hive.exec.reducers.bytes.per.reducer=<number>
In order to limit the maximum number of reducers:
  set hive.exec.reducers.max=<number>
In order to set a constant number of reducers:
  set mapreduce.job.reduces=<number>
Starting Job = job_1646476607176_0162, Tracking URL
Kill Command = /ddhome/bin/hadoop/bin/hadoop job  -
Hadoop job information for Stage-1: number of mappe
2022-03-25 10:10:21,536 Stage-1 map = 0%,  reduce =
2022-03-25 10:10:28,959 Stage-1 map = 100%,  reduce
2022-03-25 10:10:35,117 Stage-1 map = 100%,  reduce
MapReduce Total cumulative CPU time: 4 seconds 100
Ended Job = job_1646476607176_0162
MapReduce Jobs Launched:
Stage-Stage-1: Map: 1  Reduce: 1   Cumulative CPU:
Total MapReduce CPU Time Spent: 4 seconds 100 msec
OK
1002    102     10255
1002    101     9852
1001    102     9852
1003    102     9654
1001    101     8053
1001    103     7569
1002    103     6951
1003    101     6851
```

图4-33 字段sla按降序排序

我们对该员工工资表dep_id字段中的数据进行分析发现，该字段有部分数据为重复数据，那么我们考虑将同部门的数据放在一起并以降序的方式进行排序。也就是对dep_id进行升序（asc）和降序（desc）排序，Hive可以使用sort by关键字来实现，我们把它称为局部排序。执行dep_id字段并按降序的方式排序的HQL命令如下所示：

```
hive> select * from dep_sla sort by dep_id desc;
```

对dep_id字段按局部降序排序的结果如图4-34所示。

```
hive> select * from dep_sla sort by dep_id desc;
WARNING: Hive-on-MR is deprecated in Hive 2 and may
.e. spark, tez) or using Hive 1.X releases.
Query ID = hw18883508548_20220325101659_d2d78ce5-b9
Total jobs = 1
Launching Job 1 out of 1
Number of reduce tasks not specified. Estimated fro
In order to change the average load for a reducer (
  set hive.exec.reducers.bytes.per.reducer=<number>
In order to limit the maximum number of reducers:
  set hive.exec.reducers.max=<number>
In order to set a constant number of reducers:
  set mapreduce.job.reduces=<number>
Starting Job = job_1646476607176_0164, Tracking URI
Kill Command = /ddhome/bin/hadoop/bin/hadoop job  -
Hadoop job information for Stage-1: number of mappe
2022-03-25 10:17:27,107 Stage-1 map = 0%,  reduce =
2022-03-25 10:17:39,217 Stage-1 map = 100%,  reduce
2022-03-25 10:17:47,456 Stage-1 map = 100%,  reduce
MapReduce Total cumulative CPU time: 4 seconds 780
Ended Job = job_1646476607176_0164
MapReduce Jobs Launched:
Stage-Stage-1: Map: 1  Reduce: 1   Cumulative CPU:
Total MapReduce CPU Time Spent: 4 seconds 780 msec
OK
1002    103     6951
1001    103     7569
1003    102     9654
1002    102     10255
1001    102     9852
1003    101     6851
1002    101     9852
1001    101     8053
```

图4-34 对dep_id字段按局部降序排序的结果

对于该员工表中的数据，我们想对同部门员工的工资进行降序排序，使用的关键字为distribute by。通常情况下，distribute by和sort by联合使用。通过大量实验得知，想要得出正确的实验结果，在使用distribute by之前，必须对reducer的数量进行设置，即要设置reducer的数量为部门总数，该数据中，部门数为3。该过程的HQL代码如下所示：

```
hive> set mapreduce.job.reduces=3;
hive> select * from table_name  distribute by dep_id sort by sla desc;
```

对同部门的员工工资按降序排序的结果如图4-35所示。

```
hive> set mapreduce.job.reduces=3;
hive> select * from dep_sla  distribute by dep_id sort by sla desc;
WARNING: Hive-on-MR is deprecated in Hive 2 and may not be available in
.e. spark, tez) or using Hive 1.X releases.
Query ID = hw18883508548_20220325110443_f38f00bb-97b4-4bd0-99ef-1ee3ac8
Total jobs = 1
Launching Job 1 out of 1
Number of reduce tasks not specified. Defaulting to jobconf value of: 3
In order to change the average load for a reducer (in bytes):
  set hive.exec.reducers.bytes.per.reducer=<number>
In order to limit the maximum number of reducers:
  set hive.exec.reducers.max=<number>
In order to set a constant number of reducers:
  set mapreduce.job.reduces=<number>
Starting Job = job_1646476607176_0171, Tracking URL = http://ddcvb:2808
Kill Command = /ddhome/bin/hadoop/bin/hadoop job  -kill job_16464766071
Hadoop job information for Stage-1: number of mappers: 1; number of redu
2022-03-25 11:05:07,064 Stage-1 map = 0%,  reduce = 0%
2022-03-25 11:05:22,968 Stage-1 map = 100%,  reduce = 0%, Cumulative CP
2022-03-25 11:05:35,267 Stage-1 map = 100%,  reduce = 33%, Cumulative C
2022-03-25 11:05:39,391 Stage-1 map = 100%,  reduce = 100%, Cumulative
MapReduce Total cumulative CPU time: 10 seconds 60 msec
Ended Job = job_1646476607176_0171
MapReduce Jobs Launched:
Stage-Stage-1: Map: 1  Reduce: 3   Cumulative CPU: 10.06 sec   HDFS Rea
Total MapReduce CPU Time Spent: 10 seconds 60 msec
OK
1002    102     10255
1001    102     9852
1003    102     9654
1001    103     7569
1002    103     6951
1002    101     9852
1001    101     8053
1003    101     6851
```

图4-35 对同部门的员工工资按降序排序的结果

4.4.2 分组

在实际生产环境中，对海量数据分析过程中可能会涉及的部分数据进行相关计算。例如，在dep_sla表中，如果想对某个部门的工资情况进行分析计算，那么需要用到关键字group by，我们把它称为分组。在Hive查询语句中，通常group by需要与聚合函数max()、min()和avg()等联合使用。例如，我们需要求解同部门工资的平均值，在写计算命令之前，为了方便读者理解，我们可以写出显示列名称的语句。此过程的命令如下：

```
hive> set hive.cli.print.header=true;
hive> select dep_id,avg(sla) avg_sla from dep_sla
> group by dep_id;
```

求同部门的平均工资的结果如图4-36所示。

```
hive> set hive.cli.print.header=true;
hive> select dep_id,avg(sla) avg_sla from dep_sla
    > group by dep_id;
WARNING: Hive-on-MR is deprecated in Hive 2 and may
.e. spark, tez) or using Hive 1.X releases.
Query ID = hw18883508548_20220325112305_0db2945c-8c
Total jobs = 1
Launching Job 1 out of 1
Number of reduce tasks not specified. Defaulting to
In order to change the average load for a reducer (
  set hive.exec.reducers.bytes.per.reducer=<number>
In order to limit the maximum number of reducers:
  set hive.exec.reducers.max=<number>
In order to set a constant number of reducers:
  set mapreduce.job.reduces=<number>
Starting Job = job_1646476607176_0173, Tracking URI
Kill Command = /ddhome/bin/hadoop/bin/hadoop job  -
Hadoop job information for Stage-1: number of mappe
2022-03-25 11:23:33,251 Stage-1 map = 0%,  reduce =
2022-03-25 11:23:44,952 Stage-1 map = 100%,  reduce
2022-03-25 11:23:59,345 Stage-1 map = 100%,  reduce
2022-03-25 11:24:00,386 Stage-1 map = 100%,  reduce
MapReduce Total cumulative CPU time: 10 seconds 110
Ended Job = job_1646476607176_0173
MapReduce Jobs Launched:
Stage-Stage-1: Map: 1  Reduce: 3   Cumulative CPU:
Total MapReduce CPU Time Spent: 10 seconds 110 msec
OK
dep_id  avg_sla
102     9920.333333333334
103     7260.0
101     8252.0
```

图4-36 求同部门的平均工资的结果

求解出部门平均工资后，若想要一部分满足条件的数据，则可以结合hiving关键字进行限定。例如，我们要查询部门平均工资大于8000元的情况，执行语句如下：

```
hive> select dep_id,avg(sla) avg_sla from dep_sla
    > group by dep_id
> having avg_sla>8000;
```

查询出部门平均工资大于8000元的部门的结果如图4-37所示。

```
hive> select dep_id,avg(sla) avg_sla from dep_sla
    > group by dep_id
    > having avg_sla>8000;
WARNING: Hive-on-MR is deprecated in Hive 2 and may
.e. spark, tez) or using Hive 1.X releases.
Query ID = hw18883508548_20220325112912_21874fe5-97
Total jobs = 1
Launching Job 1 out of 1
Number of reduce tasks not specified. Defaulting to
In order to change the average load for a reducer (
  set hive.exec.reducers.bytes.per.reducer=<number>
In order to limit the maximum number of reducers:
  set hive.exec.reducers.max=<number>
In order to set a constant number of reducers:
  set mapreduce.job.reduces=<number>
Starting Job = job_1646476607176_0174, Tracking URI
Kill Command = /ddhome/bin/hadoop/bin/hadoop job  -
Hadoop job information for Stage-1: number of mappe
2022-03-25 11:29:39,100 Stage-1 map = 0%,  reduce =
2022-03-25 11:29:54,165 Stage-1 map = 100%,  reduce
2022-03-25 11:30:11,648 Stage-1 map = 100%,  reduce
MapReduce Total cumulative CPU time: 12 seconds 620
Ended Job = job_1646476607176_0174
MapReduce Jobs Launched:
Stage-Stage-1: Map: 1  Reduce: 3   Cumulative CPU:
Total MapReduce CPU Time Spent: 12 seconds 620 msec
OK
dep_id  avg_sla
102     9920.333333333334
101     8252.0
```

图4-37 查询出部门平均工资大于8000元的部门的结果

4.5 本章小结

本章首先介绍了Hive数据仓库的相关概念，包括Hive发展历程、Hive特点以及Hive的体系结构等内容。其次介绍了Hive安装部署流程以及配置Hive需要修改的配置文件等内容。然后介绍了Hive的基本操作（Hive常见的数据类型、常见的函数）和Hive的表操作（内部表、外部表、桶表、分区表以及表的增删改查命令的语法规则）等内容。最后介绍了Hive SQL语句的使用方法以及简单的计算等内容。

4.6 课后习题

1. 请完成以下操作，并将每一步的命令及运行结果分别以文本文件和截图的形式进行保存。

（1）创建Hive数据库：DB。

（2）查看DB数据库的描述信息。

（3）创建表cat，包含两个字段cat_id和cat_name，字符类型均为string。

（4）创建表cat2，包含两个字段cat_id和cat_name，字符类型均为string。

（5）修改cat表的表结构。在cat表中添加两个字段emp_id和cat_code，字符类型均为string。

（6）修改cat2表的表名。将cat2表重命名为cat3 。

（7）删除名为cat3的表并查看现有表。

（8）创建emp表，包含emp_id、name和age三个字段，以"\t"为分隔符。

（9）将Linux本地emp.txt文件数据导入Hive的emp表中。

（10）在emp表中，查询年龄在18岁以上的数据信息，用like或rlike查询name字段中姓名带有"N"和第二个字符为"L"的数据信息。

（11）在emp表中，以emp_id进行分组并计算平均年龄，并用having显示出平均年龄大于18的数据。

（12）用order by对年龄进行全局排序。用distribute by结合sort by对emp_id和age进行排序并设置age为降序方式。

（13）在Hive中创建一个info表，包含info_id、major和course_score三个字段，具体数据类型根据数据自行定义,以"\t"为分隔符。将Linux本地info.txt文件数据导入Hive的info表中。

（14）在info表中用insert插入如下数据：

```
1005  bigdata Math:68,hadoop:89,hive:91,hbase:86
```

（15）用join on 语句实现info表和emp表的内连接和外连接（左外连接、右外连接、全外连接）。

(16)在Hive中创建一个分区表info_par，包含info_id、info_major和info_course_score三个普通字段，字符类型为根据实际数据进行定义。其中分区字段为year，字符类型为string，以"\t"为分隔符。

(17)将Linux本地下的info.txt表中的数据插入分区表info_par中，其中分区字段year=2021。

(18)创建一个名为emp_buck的分桶表，包含emp_id、name和age三个字段，字符类型分别为string、string和int，按照age列进行分桶，划分成两个桶。

实验数据如下。

emp.txt

123_emp.id	123_emp.name	123_emp.age
1003	SCOT	16
1002	BLAKE	17
1005	FORD	18
1003	CLARK	18
1001	SMITH	18
1004	KING	19
1002	JONES	19
1005	JAMES	20
1001	ALLEN	21

info.txt

info_id	major	course_score
1001	bigdata	Math:88,hadoop:69,hive:99,hbase:87
1001	computer	Math:70,hadoop:79,hive:99,hbase:66
1002	bigdata	Math:89,hadoop:89,hive:99,hbase:69
1002	bigdata	Math:68,hadoop:83,hive:99,hbase:81
1002	computer	Math:98,hadoop:69,hive:99,hbase:82
1003	computer	Math:74,hadoop:80,hive:99,hbase:69
1003	computer	Math:90,hadoop:74,hive:99,hbase:97
1004	bigdata	Math:69,hadoop:82,hive:99,hbase:77
1004	computer	Math:83,hadoop:81,hive:99,hbase:69
1005	computer	Math:84,hadoop:71,hive:99,hbase:77

2. 学生表student、课程表course、教师表teacher、分数表score、字段名和实验数据如下：

student.s_id　student.s_name　student.s_birth　student.s_sex
01　赵雷　1990-01-01　男
02　钱电　1990-12-21　男
03　孙风　1990-05-20　男
04　李云　1990-08-06　男
05　周梅　1991-12-01　女
06　吴兰　1992-04-01　女
07　郑竹　1989-07-01　女
08　王菊　1990-01-20　女

course.c_id　course.c_course　course.t_id
01　语文　02
02　数学　01
03　英语　03

teacher.t_id　teacher.t_name
01　张三

```
02    李四
03    王五

score.s_id  score.c_id  score.s_score
01    01    80.0
01    02    90.0
01    03    99.0
02    01    70.0
02    02    60.0
02    03    80.0
03    01    80.0
03    02    80.0
03    03    80.0
04    01    50.0
04    02    30.0
04    03    20.0
05    01    76.0
05    02    87.0
06    01    31.0
06    03    34.0
07    02    89.0
```

求解：

（1）查询“01”课程比“02”课程成绩高的学生的信息及课程分数。

（2）查询“01”课程比“02”课程成绩低的学生的信息及课程分数。

（3）查询平均成绩大于等于60分的同学的学生编号、学生姓名和平均成绩。

（4）查询平均成绩小于60分的同学的学生编号、学生姓名和平均成绩。

（5）查询所有同学的学生编号、学生姓名、选课总数、所有课程的总成绩。

（6）查询“李”姓老师的数量。

（7）查询学过“张三”老师授课的同学的信息。

第5章 Flume开发应用

学习目标

（1）通过本章的学习，可让初学者了解大数据日志采集技术。

（2）由浅入深地认识Flume日志采集概念，为初学者对后续章节的学习打下基础。

（3）通过开发案例进一步了解目前主流大数据日志采集的自定义开发，了解未来主流的大数据开发技术，带领初学者走进大数据技术。

思政目标

（1）我国在“大数据生态”发展战略的大背景下，学习大数据将对各个产业链的转型起到推动作用，让学生树立全局意识，愿意投身到我国大数据技术强国信息化的建设中来。

（2）学习大数据技术，为我国民族品牌提供服务，为增强核心技术自主能力提供服务。

5.1 Flume概述

Flume是Cloudera提供的一个高可用的、高可靠的、分布式的海量日志采集、聚合和传输的系统。Flume支持在日志系统中定制各类数据发送方，用于收集各类型数据；同时，Flume支持定制各类数据接收方，用于最终存储数据。

目前，Flume是Apache的一个顶级项目。当前，Flume有两个版本：Flume 0.9X版本的统称Flume-og和Flume 1.X版本的统称Flume-ng。由于Flume-ng经过了重大重构，与Flume-og有很大不同，所以使用时请注意区分。Web服务器每天会产生大量的日志，我们要把这些日志收集起来，移动到Hadoop平台上进行分析。那么如何移动这些数据呢？一种方法是通过输入shell cp命令到Hadoop集群上，然后通过hdfs dfs

-put命令移动数据。这种方法毫无疑问是可行的，但是有一个问题，如果在移动数据的时候，一台机器宕机了，该怎么办？这种方法没有办法进行监控，但负载均衡等技术可以解决此问题，这时需要Flume的核心支持。

Flume源使用由外部源（如Web服务器）传递给它的事件。外部源以目标Flume源可识别的格式将事件发送到Flume。例如，Avro Flume源可用于从Avro客户端或流中的其他Flume代理接收Avro事件，这些代理从Avro接收器发送事件。可以使用Thrift Flume Source定义类似的流程，以接收来自Thrift Sink或Flume Thrift Rpc客户端的事件或使用Flume Thrift协议生成的任何语言编写的Thrift客户端。当Flume源接收到事件时，它会将其存储到一个或多个通道。该通道是一个被动的存储库，可以保存事件直到它被Flume接收器消耗掉。文件通道就是一个例子——由本地文件系统支持。接收器从通道中删除事件并将其放入HDFS等外部存储库（通过Flume HDFS接收器）或将其转发到流中的下一个Flume代理（下一个跃点）的Flume源。给定代理中的源代码和接收器与通道中暂存的事件异步运行。详细的数据流模型架构如图5-1所示。

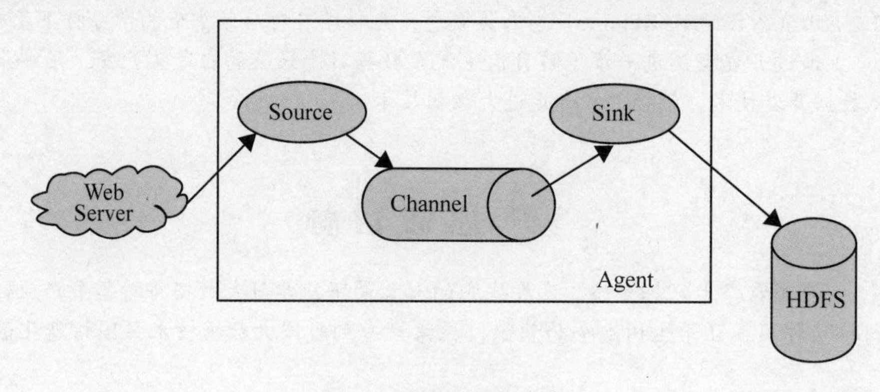

图5-1 数据流模型结构

5.2 Flume行业应用

Flume作为开源社区项目，国内有企业在使用Flume吗？答案是肯定的。开源社区同样也离不开企业开发者提供的代码。下面介绍华为公司，并深入了解Flume在企业中的应用。华为公司是全球领先的信息与通信技术解决方案供应商，专注于通信技术领域，坚持稳健经营、持续创新、开放合作。华为公司在电信运营商、企业、终端和云计算等领域构筑了端到端的解决方案优势，为运营商客户、企业客户和消费者提供有竞争力的通信技术解决方案，并致力于实现未来信息社会，构建更美好的全连接世界。从华为公司的产品、业务群体不难看出，所有客户群体都要用到日志采集系统，下面将对华为公司的日志采集系统进行详细介绍。

5.2.1 华为云日志服务

云日志服务（Log Tank Service）提供一站式日志采集、秒级搜索、海量存储、结构化处理、转储和可视化图表等功能，满足应用运维、网络日志可视化分析、等保合规和运营分析等应用场景。日志架构平台如图5-2所示。

图5-2 日志架构平台

5.2.2 企业核心集成

在企业集成项目中，华为公司除了云日志服务使用Flume作为日志采集引擎外，在华为企业级大数据平台，MRS服务的大数据MapReduce计算架构也同样使用Flume作为集成数据采集组模块。华为企业级大数据平台架构图如图5-3所示。

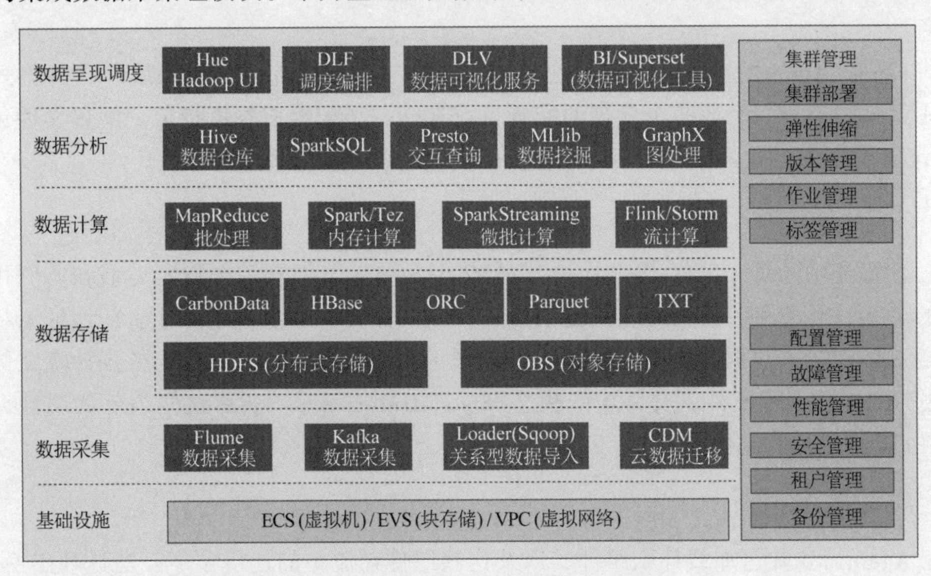

图5-3 华为企业级大数据平台架构图

通过以上内容的学习，基本了解了我国企业使用Flume的情况，我们学习的技术几乎全为社区知识文化。

5.3 安装Flume

微课：
v5-1 Flume软件包下载说明

在安装管理步骤中，flume-ng步骤的安装方法比flume-og更简单，安装数据更小，也更易于部署。安装时，请保持flume-ng与flume-og的向后兼容性。目前Flume社区正在测试flume-ng的兼容性，以及其他系统的集成与大数据应用的兼容性。

5.3.1 下载Flume源码

建议读者从Flume官方网站下载，下载时从列表中的链接显示可用镜像，根据下载时的位置默认进行选择。如果在下载过程中没有看到该页面文件，请尝试使用其他浏览器。下载链接和源文件详情如下所示：

```
#官方下载链接
https://flume.apache.org/download.html
#下载源文件
apache-flume-1.9.0-src.tar.gz
```

5.3.2 安装Agent

Flume Agent的配置存储在一个本地文件中。这是一个遵循Java属性文件格式的文本文件。对于一个或多个代理的配置可以在同一个配置文件里指定。配置文件包括在Agent中的每一个source、sink和channel的属性及其数据流连接。

1. 配置单个组件

数据流中的每个组件（source、sink或者channel）都有一个特定类型和实例化名称的名字、类型和一系列属性。比如Avro Source需要一个主机名（或者IP地址）和一个端口号来接收数据。内存通道可以有最大的队列大小，HDFS Sink需要知晓文件系统的URI地址创建文件、文件访问频率hdfs.rollInterval等。所有这些组件的属性都需要在Flume代理的文件中设置。

2. 连接各个组件

Agent需要知道加载什么组件，以及这些组件在流中的连接顺序。通过列出Agent中的source、sink和channel名称来定义每个sink与source的channel。

3. 启动Agent

bin目录下的flume-ng是Flume的启动脚本，启动时需要指定Agent的名字、配置文件的目录和配置文件的名称，如下：

```
#bin/flume-ng agent -n $agent_name -c conf -f conf/
   flume-conf.properties.template
```

4. 简单的Hello World

下面看一个配置文件的例子。配置一个单节点的Flume，这个配置是先让自己生成Event数据，然后Flume将它们输出到控制台上。在这个配置文件中，source使用的是NetCat TCP Source，简单来说就是监听本机某个端口上接收到的TCP协议的消息，收到的每行内容都会解析封装成一个Event，然后发送到channel；sink使用的是Logger Sink，这个sink可以将Event输出到控制台；channel使用的是内存channel，是一个用内存作为Event缓冲的channel。实现代码如下所示：

```
#example.conf:配置一个单节点的Flume
#配置Agent a1各个组件的名称
a1.sources = r1     #Agent a1的source有一个，叫r1
a1.sinks = k1       #Agent a1的sink也有一个，叫k1
a1.channels = c1    #Agent a1的channel有一个，叫c1
#配置Agent a1的source r1的属性
a1.sources.r1.type = netcat       #使用的是NetCat TCP Source，这里配置的是
别名，Flume内置的一些组件都是有别名的，没有别名，填写全限定类名
a1.sources.r1.bind = localhost    #NetCat TCP Source监听的主机名是本机
a1.sources.r1.port = 44444        #监听的端口
#配置Agent a1的sink k1的属性
a1.sinks.k1.type = logger         #sink使用的是Logger Sink，这里配置的也是别名
#配置Agent a1的channel c1的属性，channel是用来缓冲Event数据的
a1.channels.c1.type = memory
#channel的类型是内存channel，顾名思义，这个channel是使用内存来缓冲数据的
a1.channels.c1.capacity = 1000
#内存channel的容量大小是1000，注意这个容量不是越大越好，配置越大，一旦Flume挂掉，
丢失的event就越多
a1.channels.c1.transactionCapacity = 100
#source和sink每次从内存channel中传输的event数量
#将source和sink绑定到channel上
a1.sources.r1.channels = c1       #与source r1绑定的channel有一个，叫c1
a1.sinks.k1.channel = c1          #与sink k1绑定的channel有一个，叫c1
```

Hello World在配置文件里的注释已经写得很明白，该配置文件定义了一个Agent叫a1，a1有一个source监听本机44444端口上接收到的数据、一个缓冲数据的channel，以及一个将Event数据输出到控制台的sink。该配置文件给各个组件命名，并且设置它们的类型和其他属性。通常，一个配置文件里可能有多个Agent，当启动Flume的时候，通常会给一个Agent名字来作为程序运行的标记。用以下命令加载这个配置文件来启动Flume。

```
#bin/flume-ng agent --conf conf --conf-file example.conf --name a1
-Dflume.root.logger=INFO,console
```

我们测试一下这个例子吧！打开一个新的终端窗口，用telnet命令连接本机的44444端口，输入Hello World！后按回车，这时收到服务器的响应“OK”，说明一行数据已经成功发送。实现命令以及显示结果如下所示：

```
#telnet localhost 44444
Trying 127.0.0.1...
Connected to localhost.localdomain (127.0.0.1).
Escape character is ‘^]’.
Hello World! <ENTER>
OK
```

Flume的终端里会以log的形式输出这个收到的Event内容，详细显示结果如下所示：

```
12/06/19 15:32:19 INFO source.NetcatSource:Source starting
12/06/19 15:32:19 INFO source.NetcatSource:Created serverSocket:sun.
nio.ch.ServerSocketChannelImpl[/127.0.0.1:44444]
12/06/19 15:32:34 INFO sink.LoggerSink:Event:{headers:{} body:48 65 6C
6C 6F 20 77 6F 72 6C 64 21 0D    Hello World!.}
```

5. 基于Zookeeper的配置

Flume支持使用Zookeeper配置Agent。Zookeeper也具有实训性的功能。配置文件需要上传到Zookeeper中，并有可配置的前缀。配置文件保存在Zookeeper节点数据里。下面是a1和a2 Agent在Zookeeper下的节点树的配置信息：

```
#- /flume
 |- /a1 [Agent config file]
 |- /a2 [Agent config file]
#bin/flume-ng agent --conf conf -z zkhost:2181,zkhost1:2181 -p /flume
--name a1 -Dflume.root.logger=INFO,console
```

6. 安装第三方插件

Flume有完整的插件架构，尽管Flume已经提供了很多现成的source、channel、sink、serializer。然而，通过把自定义组件的jar包添加到flume-env.sh文件的FLUME_CLASSPATH变量中也是常有的事。现在Flume支持在一个特定的文件夹自动获取组件，这个文件夹就是plugins.d。这样使得插件的包管理、调试、错误定位更加方便，尤其是依赖包的冲突处理。plugins.d文件夹的所在位置是 $ FLUME_HOME/plugins.d，在启动时，flume-ng会启动脚本检查这个文件夹并将符合格式的插件添加到系统中。以下是两个插件的目录结构：

```
plugins.d/
plugins.d/custom-source-1/
plugins.d/custom-source-1/lib/my-source.jar
plugins.d/custom-source-1/libext/spring-core-2.5.6.jar
plugins.d/custom-source-2/
plugins.d/custom-source-2/lib/custom.jar
plugins.d/custom-source-2/native/gettext.so
```

5.3.3 数据获取

Flume支持多种从外部获取数据的方式。Flume发行版中包含的Avro客户端可以使用Avro RPC机制将给定文件发送到Flume Avro Source。执行命令如下所示：

```
#bin/flume-ng avro-client -H localhost -p 41414 -F /usr/logs/log.10
```

5.3.4 数据组合

日志收集场景中比较常见的是，数百个日志生产者发送数据到几个日志消费者Agent上，然后消费者Agent负责把数据发送到存储系统。例如，将从数百个Web服务器收集的日志发送到十几个Agent上，然后由十几个Agent写入HDFS集群。Agent架构如图5-4所示。

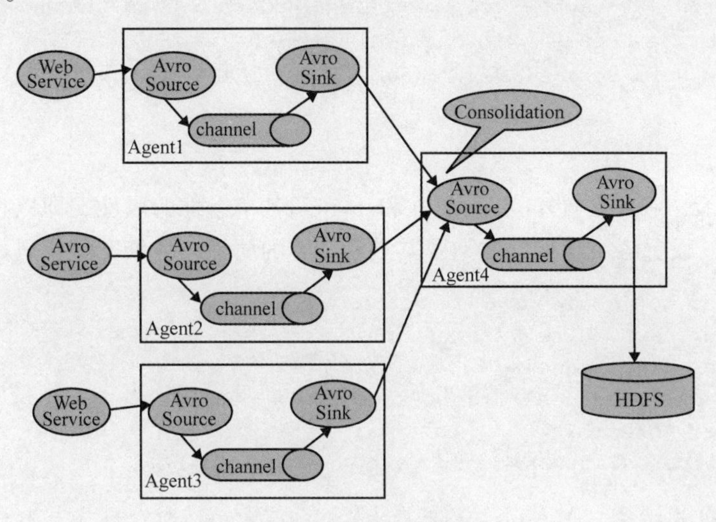

图5-4 Agent架构图

可以通过使用Avro Sink配置多个第一层Agent（Agent1、Agent2、Agent3），所有第一层Agent的Sink都指向下一级同一个Agent（Agent4）的Avro Source上，同样你也可以使用Thrift协议的Source和Sink来代替。Agent4上的Source将Event合并到一个channel中，该channel中的Event最终由HDFS Sink消费者发送到最终目的地。

5.3.5 环境配置

Flume Agent程序配置是从类似于具有分层属性设置的Java属性文件格式的文件中读取的。Flume的配置步骤如下。

1. 定义流

要在单个Agent中定义流，你需要通过channel连接source和sink。需要在配置文件中列出所有的source、sink和channel，然后将source和sink指向channel。一个source可以连接多个channel，但是sink只能连接一个channel，详细配置信息如下所示：

```
# 列出Agent的所有source、channel、sink
<Agent>.sources = <Source>
<Agent>.sinks = <Sink>
<Agent>.channels = <Channel1> <Channel2>
# 设置channel和source的关联
<Agent>.sources.<Source>.channels = <Channel1> <Channel2>
# 这一行就是给Source配置了多个channel
# 设置channel和sink的关联
<Agent>.sinks.<Sink>.channel = <Channel1>
# 列出Agent的所有source、sink和channel
agent_foo.sources = avro-appserver-src-1
agent_foo.sinks = hdfs-sink-1
agent_foo.channels = mem-channel-1
agent_foo.sources.avro-appserver-src-1.channels = mem-channel-1
# 指定与source avro-appserver-src-1相连接的channel是mem-channel-1
agent_foo.sinks.hdfs-sink-1.channel = mem-channel-1
# 指定与sink hdfs-sink-1相连接的channel是mem-channel-1
```

2.配置单个组件

定义流后，需要设置source、sink和channel各个组件的属性，即以相同的分级命名空间的方式设置各个组件的类型及其类型特有的属性。详细配置代码如下所示：

```
# properties for sources
<Agent>.sources.<Source>.<someProperty> = <someValue>
# properties for channels
<Agent>.channel.<Channel>.<someProperty> = <someValue>
# properties for sinks
<Agent>.sources.<Sink>.<someProperty> = <someValue>
```

每个组件都应该有一个type属性，这样Flume才能知道它是什么类型的组件。在前面的示例中，我们已经构建了一个以avro-appserver-src-1到hdfs-sink-1的数据流，下面继续给这几个组件配置属性。详细代码如下所示：

```
# 列出所有的组件
agent_foo.sources = avro-AppSrv-source
agent_foo.sinks = hdfs-Cluster1-sink
agent_foo.channels = mem-channel-1
# 将source、sink与channel相连接
# 配置avro-AppSrv-source的属性
agent_foo.sources.avro-AppSrv-source.type = avro
# avro-AppSrv-source的类型是Avro Source
agent_foo.sources.avro-AppSrv-source.bind = localhost
# 监听的主机名或者IP地址是localhost
agent_foo.sources.avro-AppSrv-source.port = 10000
# 监听的端口是10000
# 配置mem-channel-1的属性
agent_foo.channels.mem-channel-1.type = memory
# channel的类型是内存channel
```

```
agent_foo.channels.mem-channel-1.capacity = 1000
# channel的最大容量是1000
agent_foo.channels.mem-channel-1.transactionCapacity = 100
# source和sink每次是从channel写入和读取Event数量
# 配置hdfs-Cluster1-sink的属性
agent_foo.sinks.hdfs-Cluster1-sink.type = hdfs
# sink的类型是HDFS Sink
agent_foo.sinks.hdfs-Cluster1-sink.hdfs.path = hdfs://namenode/flume/
webdata
# 写入HDFS目录路径
```

3. 在Agent中增加一个流

Flume Agent中可以包含多个独立的流。你可以在配置文件中列出所有的source、sink和channel等组件，这些组件可以连接成多个流。可以将source、sink连接到对应的channel上来定义两个不同的流。例如想在一个Agent中配置两个流，一个流从外部Avro客户端接收数据，然后输出到外部的HDFS，另一个流从文件读取内容，然后输出到Avro Sink，详细配置代码如下所示：

```
# 列出当前配置的所有source、sink和channel
agent_foo.sources = avro-AppSrv-source1 exec-tail-source2
# 该Agent中有2个sourse，分别是avro-AppSrv-source1和exec-tail-source2
agent_foo.sinks = hdfs-Cluster1-sink1 avro-forward-sink2
# 该Agent中有2个sink，分别是hdfs-Cluster1-sink1和avro-forward-sink2
agent_foo.channels = mem-channel-1 file-channel-2
# 该Agent中有2个channel，分别是mem-channel-1 file-channel-2
# 这是第一个流的配置
agent_foo.sources.avro-AppSrv-source1.channels = mem-channel-1
# 与avro-AppSrv-source1相连接的channel是mem-channel-1
agent_foo.sinks.hdfs-Cluster1-sink1.channel = mem-channel-1
# 与hdfs-Cluster1-sink1相连接的channel是mem-channel-1
# 这是第二个流的配置
agent_foo.sources.exec-tail-source2.channels = file-channel-2
# 与exec-tail-source2相连接的channel是file-channel-2
agent_foo.sinks.avro-forward-sink2.channel = file-channel-2
# 与avro-forward-sink2相连接的channel是file-channel-2
```

4. 配置多Agent流

当配置一个多层级的流时，需要在第一层Agent的末尾使用Avro/Thrift Sink，并且指向下一层Agent的Avro/Thrift Source。这样就能将第一层Agent的Event发送到下一层的Agent了。例如使用Avro客户端定期将文件（每个Event文件）发送到本地的Event上，然后本地的Agent可以把Event发送到另一个配置了存储功能的Agent上。详细配置代码如下所示：

```
########收集Web日志的Agent配置#############
# 列出Agent的source、sink和channel
agent_foo.sources = avro-AppSrv-source
agent_foo.sinks = avro-forward-sink
agent_foo.channels = file-channel
```

```
# 将source、channel、sink连接起来组成一个流
agent_foo.sources.avro-AppSrv-source.channels = file-channel
agent_foo.sinks.avro-forward-sink.channel = file-channel
# avro-forward-sink的属性配置
agent_foo.sinks.avro-forward-sink.type = avro
agent_foo.sinks.avro-forward-sink.hostname = 10.1.1.100
agent_foo.sinks.avro-forward-sink.port = 10000
##########HDFS的Agent配置###############
# 列出Agent的source、sink和channel
agent_foo.sources = avro-collection-source
# 只有一个source叫做avro-collection-source
agent_foo.sinks = hdfs-sink
# 只有一个sink叫做hdfs-sink
agent_foo.channels = mem-channel
# 只有一个channel叫做mem-channel
# 将source、channel、sink连接起来组成一个流
agent_foo.sources.avro-collection-source.channels = mem-channel
agent_foo.sinks.hdfs-sink.channel = mem-channel
# Avro Source的属性配置
agent_foo.sources.avro-collection-source.type = avro
agent_foo.sources.avro-collection-source.bind = 10.1.1.100
agent_foo.sources.avro-collection-source.port = 10000
```

5.4 配置过滤器

微课：
v5-2 配置
过滤器介绍

Flume提供一种工具用于以配置过滤器的形式将敏感的或生成的数据注入配置中。

5.4.1 过滤器的常见用法

过滤器格式类似于Java表达式语言，但它目前还不是一个完全工作的EL表达式解析器，详细配置代码如下所示：

```
<agent_name>.configfilters = <filter_name>
<agent_name>.configfilters.<filter_name>.type = <filter_type>
<agent_name>.sources.<source_name>.parameter =
    ${<filter_name>['<key_for_sensitive_or_generated_data>']}
<agent_name>.sinks.<sink_name>.parameter =
    ${<filter_name>['<key_for_sensitive_or_generated_data>']}
<agent_name>.<component_type>.<component_name>.parameter =
    ${<filter_name>['<key_for_sensitive_or_generated_data>']}
#or
<agent_name>.<component_type>.<component_name>.parameter =
    ${<filter_name>[" <key_for_sensitive_or_generated_data>"]}
```

```
#or
<agent_name>.<component_type>.<component_name>.parameter =
    ${<filter_name>[<key_for_sensitive_or_generated_data>]}
#or
<agent_name>.<component_type>.<component_name>.parameter = some_con
    stant_data${<filter_name>[<key_for_sensitive_or_generated_data>]}
```

5.4.2 环境变量过滤器

要实现环境变量过滤器，需在配置中隐藏密码。环境变量过滤器的配置代码如下所示：

```
a1.sources = r1
a1.channels = c1
a1.configfilters = f1
a1.configfilters.f1.type = environment
a1.sources.r1.channels = c1
a1.sources.r1.type = http
a1.sources.r1.keystorePassword = ${f1['my_keystore_password']}
#会得到值Secret123
```

a1.sources.r1.keystorePassword的配置属性将获取my_keystore_password环境变量的值，运行命令如下所示：

```
#my_keystore_password=Secret123 bin/flume-ng agent --conf conf
--conf-file example.conf
```

5.4.3 外部进程配置过滤器

外部进程配置过滤器要为滚动文件接收器生成目录的一部分，将执行以获取给定键的值的命令。该命令将被称为command和key，详细配置代码如下所示：

```
a1.sources = r1
a1.channels = c1
a1.configfilters = f1
a1.configfilters.f1.type = external
a1.configfilters.f1.command = /usr/bin/passwordResolver.sh
a1.configfilters.f1.charset = UTF-8
a1.sources.r1.channels = c1
a1.sources.r1.type = http
a1.sources.r1.keystorePassword = ${f1['my_keystore_password']}
#会得到值Secret123
```

执行命令如下所示：

```
# /usr/bin/generateUniqId.sh agent_name
```

要为滚动文件接收器生成目录的一部分，详细代码如下所示：

```
a1.sources = r1
a1.channels = c1
a1.configfilters = f1

a1.configfilters.f1.type = external
a1.configfilters.f1.command = /usr/bin/generateUniqId.sh
a1.configfilters.f1.charset = UTF-8
a1.sinks = k1
a1.sinks.k1.type = file_roll
a1.sinks.k1.channel = c1
a1.sinks.k1.sink.directory = /var/log/flume/agent_${f1['agent_name']}
# 将是/var/log/flume/agent_1234
```

执行命令如下所示：

```
#/usr/bin/passwordResolver.sh my_keystore_password
```

5.4.4 Hadoop存储配置过滤器

类路径的存储功能需要hadoop-common库2.6以上的版本支持，如果安装了Hadoop，代理会自动将其添加到类路径中。属性credential.provider.path可以提供路径，属性credstore.java-keystore-provider.password如果文件用于存储密码，则为密码文件的名称。该文件必须在类路径中，提供者可以使用密码。配置代码如下所示：

```
a1.sources = r1
a1.channels = c1
a1.configfilters = f1
a1.configfilters.f1.type = hadoop
a1.configfilters.f1.credential.provider.path = jceks://file/<path_to_jceks
file>
a1.sources.r1.channels = c1
a1.sources.r1.type = http
a1.sources.r1.keystorePassword = ${f1['my_keystore_password']}
#从存储中获取值
```

5.5 Flume自定义实现

微课：
v5-3 自定义客户端

使用Flume采集服务器本地日志，需要按照日志类型的不同来将不同种类的日志发往不同的分析系统。本节将以端口数据模拟日志，以数字（单个）和字母（单个）模拟不同类型的日志，再自定义interceptor来区分数字和字母，将其分别发往不同的分析系统。

5.5.1 RPC客户端

从Flume 1.4.0开始，Avro是默认的RPC协议。NettyAvroRpcClient和ThriftRpcClient实现了RpcClient接口。客户端需要用Flume代理的主机和端口创建RpcClient接口，然后才能使用RpcClient向代理发送数据。以下代码展示了如何在用户的应用程序中使用Flume Client SDK API：

```
import org.apache.flume.Event;
import org.apache.flume.EventDeliveryException;
import org.apache.flume.api.RpcClient;
import org.apache.flume.api.RpcClientFactory;
import org.apache.flume.event.EventBuilder;
import java.nio.charset.Charset;
public class MyApp {
    public static void main(String[] args) {
        MyRpcClientFacade client = new MyRpcClientFacade();
        //Initialize client with the remote Flume agent's host and port
          client.init("host.example.org",41414);
        //Send 10 events to the remote Flume agent. That agent should be
        //configured to listen with an AvroSource.
          String sampleData = "Hello Flume!";
        for (int i = 0;i < 10; i++) {
            client.sendDataToFlume(sampleData);
        }
        client.cleanUp();
    }
}
class MyRpcClientFacade {
    private RpcClient client;
    private String hostname;
    private int port;
    public void init(String hostname,int port) {
        //Setup the RPC connection
        this.hostname = hostname;
        this.port = port;
        this.client = RpcClientFactory.getDefaultInstance(
            hostname,port);
        //Use the following method to create a thrift client (
          instead of the above line):
        //this.client = RpcClientFactory.getThriftInstance(
          hostname,port);
    }
    public void sendDataToFlume(String data) {
        //Create a Flume Event object that encapsulates the sample data
        Event event = EventBuilder.withBody(
            data,Charset.for Name("UTF-8"));
        //Send the event
        try {
            client.append(event);
        } catch (EventDeliveryException e) {
            //clean up and recreate the client
```

```
            client.close();
            client = null;
            client = RpcClientFactory.getDefaultInstance(
                hostname,port);
            //Use the following method to create a thrift client (
              in stead of the above line):
            //this.client = RpcClientFactory.getThriftInstance(
              hostname,port);
        }
    }
    public void cleanUp() {
        //Close the RPC connection
        client.close();
    }
}
```

当使用远程Flume代理时，需要一个AvroSource。如果使用的是Thrift客户端，则需要一个ThriftSource监听某个端口。以下是一个等待MyApp1连接的Flume代理配置示例：

```
a1.channels = c1
a1.sources = r1
a1.sinks = k1
a1.channels.c1.type = memory
a1.sources.r1.channels = c1
a1.sources.r1.type = avro
# For using a thrift source set the following instead of the above line.
# a1.source.r1.type = thrift
a1.sources.r1.bind = 0.0.0.0
a1.sources.r1.port = 41414
a1.sinks.k1.channel = c1
a1.sinks.k1.type = logger
```

为了获得更大的灵活性，可以使用以下属性配置默认的Flume客户端来实现NettyAvroRpcClient和ThriftRpcClient，代码如下所示：

```
client.type = default (for avro) or thrift (for thrift)
hosts = h1                           # default client accepts only 1 host
                                     # (additional hosts will be ignored)
hosts.h1 = host1.example.org:41414   # host and port must both be specified
                                     # (neither has a default)
batch-size = 100                     # Must be >=1 (default: 100)
connect-timeout = 20000              # Must be >=1000 (default: 20000)
request-timeout = 20000              # Must be >=1000 (default: 20000)
```

5.5.2 安全RPC客户端

从Flume 1.6.0开始，Thrift源和接收器支持基于Kerberos的身份验证。客户端使用SecureRpcClientFactory的getThriftInstance方法来获取SecureThriftRpcClient。Flume通过SecureThriftRpcClient扩展实现RpcClient接口的ThriftRpcClient。当程序调

用RPC客户端的SecureRpcClientFactory时，Kerberos身份验证模块位于类路径的flume-ng-auth模块中。客户端主体和客户端密钥表作为参数通过属性传入，用来反映客户端的凭据以针对Kerberos KDC进行身份验证。此外，还应提供客户端连接到目标Thrift源的服务器主体。以下代码显示了如何在用户的应用程序中使用SecureRpcClientFactory：

```
import org.apache.flume.Event;
import org.apache.flume.EventDeliveryException;
import org.apache.flume.event.EventBuilder;
import org.apache.flume.api.SecureRpcClientFactory;
import org.apache.flume.api.RpcClientConfigurationConstants;
import org.apache.flume.api.RpcClient;
import java.nio.charset.Charset;
import java.util.Properties;
public class MyApp {
  public static void main(String[] args) {
    MySecureRpcClientFacade client = new MySecureRpcClientFacade();
    //Initialize client with the remote Flume agent's host,port
    Properties props = new Properties();
    props.setProperty(RpcClientConfigurationConstants.CONFIG_
        CLIENT_TYPE,"thrift");
    props.setProperty("hosts","h1");
    props.setProperty("hosts.h1","client.example.org"+":"+
        String.valueOf(41414));
    //Initialize client with the kerberos authentication related
      properties
    props.setProperty("kerberos","true");
    props.setProperty("client-principal","flumeclient/
        client.example.org@EXAMPLE.ORG");
    props.setProperty("client-keytab","/tmp/flumeclient.keytab");
    props.setProperty("server-principal","flume/
        server.example.org@EXAMPLE.ORG");
    client.init(props);
    //Send 10 events to the remote Flume agent.That agent should be
    //configured to listen with an AvroSource.
    String sampleData = "Hello Flume!";
    for (int i = 0; i < 10; i++) {
      client.sendDataToFlume(sampleData);
    }
    client.cleanUp();
  }
}
class MySecureRpcClientFacade {
  private RpcClient client;
  private Properties properties;
  public void init(Properties properties) {
    //Setup the RPC connection
    this.properties = properties;
    //Create the ThriftSecureRpcClient instance by using
      SecureRpcClientFactory
    this.client = SecureRpcClientFactory.getThriftInstance(properties);
```

```
  }
  public void sendDataToFlume(String data) {
    //Create a Flume Event object that encapsulates the sample data
    Event event = EventBuilder.withBody(data,
        Charset.forName("UTF-8"));

    //Send the event
    try {
      client.append(event);
    } catch (EventDeliveryException e) {
      //clean up and recreate the client
      client.close();
      client = null;
      client = SecureRpcClientFactory.getThriftInstance(properties);
    }
  }

  public void cleanUp() {
    //Close the RPC connection
    client.close();
  }
}
```

远程ThriftSource应该以Kerberos模式启动。下面是一个等待MyApp2连接的Flume代理配置示例：

```
a1.channels = c1
a1.sources = r1
a1.sinks = k1
a1.channels.c1.type = memory
a1.sources.r1.channels = c1
a1.sources.r1.type = thrift
a1.sources.r1.bind = 0.0.0.0
a1.sources.r1.port = 41414
a1.sources.r1.kerberos = true
a1.sources.r1.agent-principal = flume/server.example.org@EXAMPLE.ORG
a1.sources.r1.agent-keytab = /tmp/flume.keytab
a1.sinks.k1.channel = c1
a1.sinks.k1.type = logger
```

5.5.3 故障转移客户端

故障转移客户端封装了默认的Avro RPC客户端，可以向客户端提供故障转移处理能力。这需要一个以空格分隔的host:port列表，表示构成故障转移的Flume代理。故障转移RPC客户端当前不支持Thrift。如果与当前选定的主机即代理存在通信错误，则故障转移客户端会自动将故障转移到列表的下一个主机程序，代码如下所示：

```
//Setup properties for the failover
Properties props = new Properties();
props.put("client.type","default_failover");
//List of hosts (space-separated list of user-chosen host aliases)
```

```
props.put("hosts","h1 h2 h3");
//host/port pair for each host alias
String host1 = "host1.example.org:41414";
String host2 = "host2.example.org:41414";
String host3 = "host3.example.org:41414";
props.put("hosts.h1",host1);
props.put("hosts.h2",host2);
props.put("hosts.h3",host3);
//create the client with failover properties
RpcClient client = RpcClientFactory.getInstance(props);
```

为了获得更大的灵活性，可以使用以下属性实现故障转移Flume客户端的功能，代码如下所示：

```
client.type = default_failover
hosts = h1 h2 h3                     # at least one is required,but 2 or
# more makes better sense
hosts.h1 = host1.example.org:41414
hosts.h2 = host2.example.org:41414
hosts.h3 = host3.example.org:41414
max-attempts = 3                     # Must be >=0 (default:number of hosts
# specified,3 in this case).A '0'
# value doesn't make much sense because
# it will just cause an append call to
# immmediately fail.A '1' value means
# that the failover client will try only
# once to send the Event,and if it
# fails then there will be no failover
# to a second client,so this value
# causes the failover client to
# degenerate into just a default client.
# It makes sense to set this value to at
# least the number of hosts that you
# specified.
batch-size = 100                      # Must be >=1 (default:100)
connect-timeout = 20000               # Must be >=1000 (default:20000)
request-timeout = 20000               # Must be >=1000 (default:20000)
```

5.5.4 负载均衡RPC客户端

Flume Client SDK还支持在多个主机之间进行负载均衡的RpcClient。这种类型的客户端采用空格分隔的host:port列表，表示构成负载均衡的Flume代理。这个客户端可以配置一个负载均衡策略，随机选择一个配置的主机，或者以循环的方式选择一个主机。还可以自己实现LoadBalancingRpcClient$HostSelector接口的自定义类，以便使用自定义选择顺序。这种情况下，需要将自定义类的FQCN指定为host-selector属性的值。LoadBalancing RPC客户端当前不支持Thrift。如果启用了回退，则客户端暂时将失败的主机列入黑名单，当主机被排除在被选为故障转移主机之前，直到给定的超时。当超时过去时，如果主机仍然没有响应，则认为这是一个连续故障，超时会以指数方式增加，这样可以避免在无响应的主机上陷入长时间等待。可以通过设置maxBackoff（以毫秒为单位）来配置最大退避时间。在OrderSelector类中指定maxBackoff的默认值为30秒，该类是两种负载均衡策略的超类。退避超时将随着每个连续故障而呈指数增长，直至最大可能的退避超时。代码如下所示：

```
//Setup properties for the load balancing
Properties props = new Properties();
props.put("client.type","default_loadbalance");
//List of hosts (space-separated list of user-chosen host aliases)
props.put("hosts","h1 h2 h3");
//host/port pair for each host alias
String host1 = "host1.example.org:41414";
String host2 = "host2.example.org:41414";
String host3 = "host3.example.org:41414";
props.put("hosts.h1",host1);
props.put("hosts.h2",host2);
props.put("hosts.h3",host3);
props.put("host-selector","random"); //For random host selection
//props.put("host-selector","round_robin"); //For round-robin host
//selection
props.put("backoff","true"); //Disabled by default.
props.put("maxBackoff","10000"); //Defaults 0,which effectively
//becomes 30000 ms
//Create the client with load balancing properties
RpcClient client = RpcClientFactory.getInstance(props);
```

为了获得更大的灵活性，可以使用以下属性配置负载均衡Flume客户端来实现LoadBalancingRpcClient，代码如下所示：

```
client.type = default_loadbalance
hosts = h1 h2 h3                      # At least 2 hosts are required
hosts.h1 = host1.example.org:41414
hosts.h2 = host2.example.org:41414
hosts.h3 = host3.example.org:41414
backoff = false                       # Specifies whether the client should
                                      # back-off from (i.e. temporarily
                                      # blacklist) a failed host
                                      # (default: false).
maxBackoff = 0                        # Max timeout in millis that a will
                                      # remain inactive due to a previous
                                      # failure with that host (default: 0,
                                      # which effectively becomes 30000)
host-selector = round_robin           # The host selection strategy used
batch-size = 100                      # Must be >=1 (default:100)
connect-timeout = 20000               # Must be >=1000 (default:20000)
request-timeout = 20000               # Must be >=1000 (default:20000)
```

5.5.5 Transaction接口

Transaction接口是基于Flume的稳定性考虑的。所有主要组件（即Source、Sink和Channel）都必须使用Transaction接口。也可以理解为Transaction接口就是Flume的事务，Source和Sink的发送数据与接收数据都是在Transaction里完成的。Transaction接口流程图如图5-5所示。

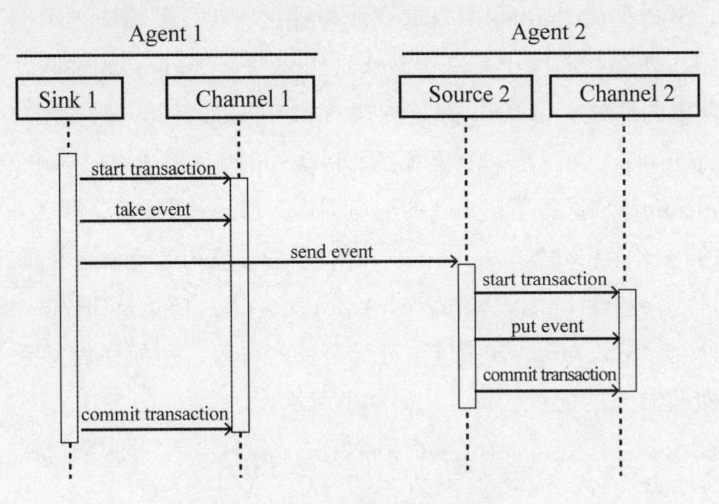

图5-5 Transaction接口流程图

从图5-5可以看出，Transaction在Channel中实现，每个连接到Channel的Source和Sink都必须获取一个Transaction对象。Source使用ChannelProcessor接口来封装Transaction，Sink通过配置的Channel显式管理它们。存放事件到Channel和从Channel中提取事件的操作是在活动的Transaction中执行的。代码如下所示：

```
Channel ch = new MemoryChannel();
Transaction txn = ch.getTransaction();
txn.begin();
try {
  //This try clause includes whatever Channel operations you want to do
  Event eventToStage = EventBuilder.withBody("Hello Flume!",
                     Charset.forName("UTF-8"));
  ch.put(eventToStage);
  // Event takenEvent = ch.take();
  // ...
  txn.commit();
} catch (Throwable t) {
  txn.rollback();
  //Log exception,handle individual exceptions as needed
  //re-throw all Errors
  if (t instanceof Error) {
    throw (Error)t;
  }
} finally {
  txn.close();
}
```

从Channel中获取了一个Transaction。在begin()返回后，事务处于活动状态/打开状态，然后将事件放入Channel中。如果put成功，则Transaction提交并关闭。

5.5.6 Sink

编程时，Sink是从Channel中提取事件并将它们转发到流中的下一个Flume代理或将它们存储在外部存储库中。一个Sink只能与一个Channel相关联，正如Flume属性文件中所配置的那样。每个Sink都有一个SinkRunner实例，当使用Flume框架调用SinkRunner.start()时，会创建一个新线程来驱动Sink调用SinkRunner.PollingRunner作为线程的Runnable。该线程管理Sink的生命周期。Flume需要实现作为LifecycleAware接口一部分的start()和stop()方法。Sink.start()方法应该初始化Sink并将其带到可以将Event转发到其下一个目的地的状态。Sink.process()方法应该从Channel中提取事件并转发其核心处理。Sink.stop()方法应该进行必要的清理，例如释放资源。Sink还需要Configurable处理自己的配置接口，代码如下所示：

```
public class MySink extends AbstractSink implements Configurable {
  private String myProp;
  @Override
  public void configure(Context context) {
    String myProp = context.getString("myProp","defaultValue");
    //Process the myProp value (e.g.validation)
    //Store myProp for later retrieval by process() method
    this.myProp = myProp;
  }
  @Override
  public void start() {
    //Initialize the connection to the external repository (e.g. HDFS)
 that
    //this Sink will forward Events to..
  }
  @Override
  public void stop() {
    //Disconnect from the external respository and do any
    //additional cleanup (e.g. releasing resources or nulling-out
    //field values)...
  }
  @Override
  public Status process() throws EventDeliveryException {
    Status status = null;
    //Start transaction
    Channel ch = getChannel();
    Transaction txn = ch.getTransaction();
    txn.begin();
    try {
      //This try clause includes whatever Channel operations you want
        to do
      Event event = ch.take();
      //Send the Event to the external repository.
      //storeSomeData(e);
      txn.commit();
```

```
      status = Status.READY;
    } catch (Throwable t) {
      txn.rollback();
      //Log exception, handle individual exceptions as needed
      status = Status.BACKOFF;
      //re-throw all Errors
      if (t instanceof Error) {
        throw (Error)t;
      }
    }
    return status;
  }
}
```

5.5.7 Source

Source是从外部客户端接收数据并将其存储到配置的Channel中。Source可以获取自己的ChannelProcessor实例来处理事件，在Channel本地事务中以串行方式提交。在发生异常的情况下，所需的Channel将抛出异常，所有Channel将回滚其事务，但之前在其他Channel上处理的事件将保持提交。与SinkRunner.PollingRunner Runnable类似，PollingRunner Runnable在Flume框架调用PollableSourceRunner.start()创建的线程上执行。每个PollableSource都与它自己运行的PollingRunner线程相关联。该线程管理PollableSource的生命周期，例如启动和停止。PollableSource必须在LifecycleAware接口中声明start()和stop()方法。PollableSource的运行器调用Source的process()方法。process()方法应该检查新数据并将其作为Flume Event存储到Channel中。实际上，有两种类型的Source提到了PollableSource。EventDrivenSource与PollableSource不同，EventDrivenSource必须有自己的回调机制来捕获新数据并将其存储到Channel中，不像PollableSource那样都由自己的线程驱动。下面是一个自定义PollableSource的示例程序代码：

```
public class MySource extends AbstractSource implements Configurable,
PollableSource {
  private String myProp;
  @Override
  public void configure(Context context) {
    String myProp = context.getString("myProp","defaultValue");
    //Process the myProp value (e.g. validation,convert to another
      type,...)
    //Store myProp for later retrieval by process() method
    this.myProp = myProp;
  }
  @Override
  public void start() {
    //Initialize the connection to the external client
  }
```

```
  @Override
  public void stop () {
    //Disconnect from external client and do any additional cleanup
    //(e.g. releasing resources or nulling-out field values)..
  }
  @Override
  public Status process() throws EventDeliveryException {
    Status status = null;
    try {
      //This try clause includes whatever Channel/Event operations you
        want to do
      //Receive new data
      Event e = getSomeData();
      //Store the Event into this Source's associated Channel(s)
      getChannelProcessor().processEvent(e);
      status = Status.READY;
    } catch (Throwable t) {
      //Log exception,handle individual exceptions as needed
      status = Status.BACKOFF;
      //re-throw all Errors
      if (t instanceof Error) {
        throw (Error)t;
      }
    } finally {
      txn.close();
    }
    return status;
  }
}
```

5.6 本章小结

通过本章的学习，我们可以了解目前主流的日志采集技术，学习分布式高可靠、高可用的服务，用于高效地收集、聚合、移动大量的日志数据。本章详细介绍了Flume的功能、Flume应用实施和环境部署安装。通过项目案例进行自定义开发。

5.7 课后习题

1.选择题

（1）Flume软件最初的开发公司是（ ）。

A.Microsoft B.Cloudera C.Google D.baidu

（2）以下选项不是Flume项目特性的是（ ）。

A.高可用的 B.高可靠的 C.分布式的 D.实时计算的

（3）以下选项不为Flume Agent中包含组件的是（ ）。

A.Source B.Sink C.Rabbit MQ D.Channel

（4）以下选项不为Flume 提供配置过滤器的是（ ）。

A.环境变量过滤器 B.外部进程配置过滤器

C.Hadoop 存储配置过滤器 D.网络连接过滤器

（5）以下选项不为Flume自定义客户端的是（ ）。

A.DHCP客户端 B.RPC客户端

C.安全RPC客户端 D.故障转移客户端

2.简答题

（1）简述Flume的作用及其特点。

（2）简述Flume安全RPC客户端。

（3）简述Flume负载均衡RPC客户端。

第6章 Kafka开发应用

学习目标

（1）通过本章的学习，可以让初学者了解开源流处理技术，再由浅入深地理解流处理技术的概念。

（2）学习Kafka技术，可为初学者学习流处理技术打下基础，为初学者进一步了解大数据流开源技术提供帮助。

（3）通过从基础安装配置到程序开发入门的学习，可以带领初学者走进流处理技术。

思政目标

（1）大数据技术源于企业，再奉献至开源社区。同样，企业赋予了每个岗位相应的责任和权利，并把企业发展的希望寄托在这些岗位上。刻苦学习，服务企业和奉献国家。

（2）学习开源社区的核心技术，深刻理解奉献精神,为中华民族的伟大复兴做出贡献。

6.1 Kafka概述

6.1.1 Kafka简介

Kafka是由Apache软件基金会开发的一个开源流处理平台，由Scala语言和Java语言编写。Kafka是一个高吞吐量的分布式发布-订阅消息系统，可以处理消费者在网站中的所有动作流数据。这种动作（如网页浏览、搜索与其他用户的行动）是在现代网络上许多社会功能的一个关键因素。这些数据通常根据吞吐量的要求来处理日志和日志聚合。对于像Hadoop一样的日志数据和离线分析系统，由于受实时处理的限制，所以这是一种可行的解决方案。Kafka的目的是通过Hadoop的并行加载机制来统一处理线上和离线的消息，通过集群来提供实时的消息。架构师Jay Kreps这样介绍Kafka名称的由来，由于自己非常喜欢Franz Kafka，并且觉得Kafka这个名字很酷，因此取了一个与消息传递系统完全不相干的名称Kafka，该名字并没有特别的含义。

Kafka的诞生是为了解决Linkedin的数据管道问题。一开始，Linkedin采用ActiveMQ来进行数据交换，2010年前后，ActiveMQ还远远无法满足Linkedin对数据传递系统的要求，经常由于各种缺陷而导致消息阻塞或者服务无法正常访问。为了解决这个问题，Linkedin的首席架构师Jay Kreps便开始组建团队进行消息传递系统的研发；Kafka可以通过I/O的磁盘数据结构提供消息的持久化，这种结构即使以TB的消息存储，也能够保持长时间的稳定性能。Kafka高吞吐量的特性，即使是非常普通的硬件，Kafka也可以支持每秒传递数百万条消息。支持通过Kafka服务器和消费机集群来区分消息，支持Hadoop并行数据加载。Kafka通过官网发布了最新版本2.5.0。

6.1.2 Kafka企业聚能

Kafka作为社区的大数据项目，也是一个优秀的流处理平台。前面介绍了华为公司在应用Flume、Kafka开发自己的产品，本章重点介绍腾讯公司同样使用Apache Kafka来开发自己的产品腾讯CKafka（Cloud Kafka）。腾讯CKafka（Cloud Kafka）是基于开源Apache Kafka消息队列引擎，提供高吞吐性能、高可扩展性的消息队列服务。消息队列CKafka完美兼容Apache Kafka。在性能、扩展性、业务安全保障、运维等方面具有超强优势，让你在享受低成本、超强功能的同时，免除烦琐运维工作。产品详细功能介绍如下。

1. 收发解耦

有效解耦生产者、消费者之间的关系。在确保同样接口约束的前提下，允许独立扩展或修改生产者/消费者之间的处理过程。

2. 削峰填谷

消息队列CKafka能够抵挡突增的访问压力，不会因为突发的超负荷的请求而完全崩溃，有效提升了系统的健壮性。

3. 顺序读/写

消息队列CKafka能够保证Partition内消息的有序性。和大部分的消息队列一致，消息队列CKafka可以保证数据按照顺序进行处理，极大提升了磁盘的效率。

4. 异步通信

在业务不需要立即处理消息的场景下，消息队列CKafka提供了消息的异步处理机制，访问量高时仅将消息放入队列中，在访问量降低后再对消息进行处理，可以缓解系统的压力。

通过对腾讯CKafka的介绍，可让读者理解目前我国企业内部的一些核心技术全部走的是自主化道路。我们在学习的过程中，同样也尽可能地走开源道路，在开源社区环境下学习能够快速成长。前面章节也提到，现在开源社区核心技术的贡献离不开

中国企业的大力支持。

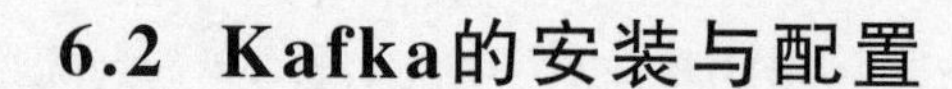

6.2 Kafka的安装与配置

安装Kafka时，我们应以官方推荐的版本为准，一般一年内会更新多个版本，社区活跃度高，发现漏洞后会很快更新。

6.2.1 资源包下载

在下载软件包的过程中，推荐从Kafka官方社区网站下载。以下为官方下载链接：

```
http://kafka.apache.org/downloads
```

6.2.2 集群环境

集群环境部署需要使用三台服务器，网络环境为内部私有网络。当遇到相关实训包无法下载时，可以通过其他网络下载，再上传至服务器中。建议关闭服务器防火墙和SELinux。关闭防火墙和SELinux的操作步骤此处省略，有兴趣的读者请参考相关内容。

6.2.3 支持软件安装

1. JDK安装

微课：
v6-1 Kafka集群配置介绍

JDK安装可去官方社区下载，本章使用的jdk软件包为jdk-8u191-linux-x64.tar.gz，下载好jdk软件包后，进入软件包目录开始安装，执行命令如下所示：

```
[root@server1 ~]# tar -zxvf jdk-8u191-linux-x64.tar.gz
jdk1.8.0_191/jre/lib/amd64/libunpack.so
jdk1.8.0_191/jre/lib/amd64/libgstreamer-lite.so
jdk1.8.0_191/jre/lib/amd64/libawt_headless.so
jdk1.8.0_191/jre/lib/amd64/libsplashscreen.so
jdk1.8.0_191/jre/lib/fontconfig.properties.src
jdk1.8.0_191/jre/lib/psfont.properties.ja
jdk1.8.0_191/jre/lib/fontconfig.Turbo.properties.src
jdk1.8.0_191/jre/lib/jce.jar
jdk1.8.0_191/jre/lib/flavormap.properties
```

```
jdk1.8.0_191/jre/lib/jfxswt.jar
jdk1.8.0_191/jre/lib/fontconfig.SuSE.10.properties.src
jdk1.8.0_191/jre/lib/fontconfig.SuSE.11.bfc
jdk1.8.0_191/jre/COPYRIGHT
jdk1.8.0_191/jre/THIRDPARTYLICENSEREADME-JAVAFX.txt
jdk1.8.0_191/jre/Welcome.html
jdk1.8.0_191/jre/README
jdk1.8.0_191/README.html
#复制jdk至系统安装目录中
[root@server1 ~]# cp -r jdk1.8.0_191 /usr/local/jdk1.8
#验证文件
[root@server1 ~]# ll /usr/local/jdk1.8/
total 25976
drwxr-xr-x. 2 root root     4096 Mar 26 23:54 bin
-r--r--r--. 1 root root     3244 Mar 26 23:54 COPYRIGHT
drwxr-xr-x. 3 root root      132 Mar 26 23:54 include
-rw-r--r--. 1 root root  5207154 Mar 26 23:54 javafx-src.zip
drwxr-xr-x. 5 root root      185 Mar 26 23:54 jre
drwxr-xr-x. 5 root root      245 Mar 26 23:54 lib
-r--r--r--. 1 root root       40 Mar 26 23:54 LICENSE
drwxr-xr-x. 4 root root       47 Mar 26 23:54 man
-r--r--r--. 1 root root      159 Mar 26 23:54 README.html
-rw-r--r--. 1 root root      424 Mar 26 23:54 release
-rw-r--r--. 1 root root 21101479 Mar 26 23:54 src.zip
-rw-r--r--. 1 root root   108062 Mar 26 23:54
THIRDPARTYLICENSEREADME-JAVAFX.txt
-r--r--r--. 1 root root   155003 Mar 26 23:54
THIRDPARTYLICENSEREADME.txt
#配置环境变量
[root@server1 ~]# vi /etc/profile
export JAVA_HOME=/usr/local/jdk1.8/
export PATH=$PATH:$JAVA_HOME/bin:$JAVA_HOME/jre/bin:$PATH
export CLASSPATH=.:$JAVA_HOME/lib:$JAVA_HOME/jre/lib
[root@server1 ~]# source /etc/profile
#验证Java版本
[root@server1 ~]# java -version
java version "1.8.0_191"
Java(TM) SE Runtime Environment (build 1.8.0_191-b12)
Java HotSpot(TM) 64-Bit Server VM (build 25.191-b12, mixed mode)
```

以上配置需对server2、server3进行相同配置，首先在server1中将jdk软件包和配置文件使用scp远程复制到另外两台服务器，再对另外两台服务器进行测试，代码如下所示：

```
#远程复制jdk软件包
[root@server1 ~]# scp -r /usr/local/jdk1.8 root@192.168.13.110:/usr/
local/
[root@server1 ~]# scp -r /usr/local/jdk1.8 root@192.168.13.120:/usr/
local/
fontconfig.Turbo.properties.src                100% 9192     3.1MB/s   00:00
jce.jar                                        100%  113KB  13.0MB/s   00:00
flavormap.properties                           100% 3901     1.7MB/s   00:00
```

```
jfxswt.jar                              100%   33KB   6.5MB/s   00:00
fontconfig.SuSE.10.properties.src       100%   16KB   5.0MB/s   00:00
fontconfig.SuSE.11.bfc                  100% 7032     2.7MB/s   00:00
COPYRIGHT                               100% 3244     1.2MB/s   00:00
THIRDPARTYLICENSEREADME-JAVAFX.txt      100%  106KB   8.4MB/s   00:00
Welcome.html                            100%  955   463.2KB/s   00:00
README                                  100%   46    18.0KB/s   00:00
README.html                             100%  159    70.8KB/s   00:00
#远程复制配置文件
[root@server1 ~]# scp -r /etc/profile root@192.168.13.110:/etc/profile
root@192.168.13.120's password:
profile                                 100% 1984   428.5KB/s   00:00
[root@server1 ~]# scp -r /etc/profile root@192.168.13.120:/etc/profile
root@192.168.13.120's password:
profile                                 100% 1984   428.5KB/s   00:00
#在server2服务器中验证
[root@server2 ~]# source /etc/profile
[root@server2 ~]# java -version
java version "1.8.0_191"
Java(TM) SE Runtime Environment (build 1.8.0_191-b12)
Java HotSpot(TM) 64-Bit Server VM (build 25.191-b12, mixed mode)
#在server3服务器中验证
[root@server3 ~]# source /etc/profile
[root@server3 ~]# java -version
java version "1.8.0_191"
Java(TM) SE Runtime Environment (build 1.8.0_191-b12)
Java HotSpot(TM) 64-Bit Server VM (build 25.191-b12, mixed mode)
```

2. Zookeeper安装

微课：v6-2 基础Zookeeper安装

为了配合Kafka实现集群部署，这里只介绍Zookeeper的安装部署过程，不介绍原理架构。本节主要介绍集群式部署知识点，分别在三台服务器中完成部署。Zookeeper安装软件包推荐在官方社区下载，下载链接如下。

```
https://dlcdn.apache.org/zookeeper/zookeeper-3.7.0/apache-zookeeper-3.7.0-bin.tar.gz
```

首先将软件包上传至server1中进行详细配置。配置好server1之后，再将软件包和配置文件通过scp远程复制到另外两台服务器，详细代码如下所示：

```
[root@server1 ~]# tar -zxvf apache-zookeeper-3.7.0-bin.tar.gz
apache-zookeeper-3.7.0-bin/lib/jackson-databind-2.10.5.1.jar
apache-zookeeper-3.7.0-bin/lib/jackson-annotations-2.10.5.jar
apache-zookeeper-3.7.0-bin/lib/jackson-core-2.10.5.jar
apache-zookeeper-3.7.0-bin/lib/jline-2.14.6.jar
apache-zookeeper-3.7.0-bin/lib/metrics-core-4.1.12.1.jar
apache-zookeeper-3.7.0-bin/lib/snappy-java-1.1.7.7.jar
[root@server1 ~]# mv apache-zookeeper-3.7.0-bin /usr/local/zookeeper
[root@server1 ~]# cd /usr/local/zookeeper/
#新建存储目录
```

```
[root@server1 zookeeper]# mkdir data
#增加配置文件
[root@server1 zookeeper]# cd conf/
[root@server1 conf]# cp zoo_sample.cfg zoo.cfg
#修改配置文件，先把dataDir=/tmp/zookeeper注释掉，然后添加以下核心配置
[root@server1 conf]# vi zoo.cfg
#============dataconfig===================
dataDir=/usr/local/zookeeper/data
server.1=192.168.13.100:2888:3888
server.2=192.168.13.110:2888:3888
server.3=192.168.13.120:2888:3888
#创建myid文件
[root@server1 conf]# cd ../data/
[root@server1 data]# touch myid
[root@server1 data]# echo "1" >> myid
#配置iptables端口
[root@server1 data]# /sbin/iptables -I INPUT -p tcp --dport 2181 -j ACCEPT
[root@server1 data]# /sbin/iptables -I INPUT -p tcp --dport 2888 -j ACCEPT
[root@server1 data]# /sbin/iptables -I INPUT -p tcp --dport 3888 -j ACCEPT
[root@server1 data]# iptables-save
# Generated by iptables-save v1.4.21 on Sun Mar 27 02:47:06 2022
*filter
:INPUT ACCEPT [24:1992]
:FORWARD ACCEPT [0:0]
:OUTPUT ACCEPT [17:1976]
-A INPUT -p tcp -m tcp --dport 3888 -j ACCEPT
-A INPUT -p tcp -m tcp --dport 2888 -j ACCEPT
-A INPUT -p tcp -m tcp --dport 2181 -j ACCEPT
#验证iptables配置
[root@server1 data]# /sbin/iptables -L -n
Chain INPUT (policy ACCEPT)
target     prot opt source               destination
ACCEPT     tcp  --  0.0.0.0/0            0.0.0.0/0            tcp dpt:3888
ACCEPT     tcp  --  0.0.0.0/0            0.0.0.0/0            tcp dpt:2888
ACCEPT     tcp  --  0.0.0.0/0            0.0.0.0/0            tcp dpt:2181
Chain FORWARD (policy ACCEPT)
target     prot opt source               destination
Chain OUTPUT (policy ACCEPT)
target     prot opt source               destination
```

在server1中配置好Zookeeper，通过scp将目录远程复制到其他两台服务器。需对另外两台服务器修改myid文件号，开放iptables三个端口，详细代码如下所示：

```
#远程复制软件包和配置文件到server2中
[root@server1 ~]# scp -r /usr/local/zookeeper root@192.168.13.110:/usr/local/
zkCli.sh                    100% 1620   286.6KB/s   00:00
zkCli.cmd                   100% 1158   274.9KB/s   00:00
zkCleanup.sh                100% 2066   369.7KB/s   00:00
README.txt                  100%  232    23.1KB/s   00:00
myid                        100%    2     0.6KB/s   00:00
#远程复制软件包和配置文件到server3中
[root@server1 ~]# scp -r /usr/local/zookeeper root@192.168.13.120:/usr/local/
```

```
zkCli.sh                      100% 1620   286.6KB/s   00:00
zkCli.cmd                     100% 1158   274.9KB/s   00:00
zkCleanup.sh                  100% 2066   369.7KB/s   00:00
README.txt                    100%  232    23.1KB/s   00:00
myid                          100%    2     0.6KB/s   00:00
#修改server2配置
[root@server2 ~]# cd /usr/local/zookeeper/data/
[root@server2 data]# echo "2" > myid
[root@server2 data]# cat myid
2
[root@server2 ~]# /sbin/iptables -I INPUT -p tcp --dport 2181 -j ACCEPT
[root@server2 ~]# /sbin/iptables -I INPUT -p tcp --dport 2888 -j ACCEPT
[root@server2 ~]# /sbin/iptables -I INPUT -p tcp --dport 3888 -j ACCEPT
[root@server2 ~]# iptables-save
# Generated by iptables-save v1.4.21 on Sun Mar 27 04:27:48 2022
*filter
:INPUT ACCEPT [14:1144]
:FORWARD ACCEPT [0:0]
:OUTPUT ACCEPT [9:952]
-A INPUT -p tcp -m tcp --dport 3888 -j ACCEPT
-A INPUT -p tcp -m tcp --dport 2888 -j ACCEPT
-A INPUT -p tcp -m tcp --dport 2181 -j ACCEPT
COMMIT
# Completed on Sun Mar 27 04:27:48 2022
#修改server3配置
[root@server3 ~]# cd /usr/local/zookeeper/data/
[root@server3 data]# echo "3" > myid
[root@server3 data]# cat myid
3
[root@server3 ~]# /sbin/iptables -I INPUT -p tcp --dport 2181 -j ACCEPT
[root@server3 ~]# /sbin/iptables -I INPUT -p tcp --dport 2888 -j ACCEPT
[root@server3 ~]# /sbin/iptables -I INPUT -p tcp --dport 3888 -j ACCEPT
[root@server3 ~]# iptables-save
# Generated by iptables-save v1.4.21 on Sun Mar 27 04:27:48 2022
*filter
:INPUT ACCEPT [14:1144]
:FORWARD ACCEPT [0:0]
:OUTPUT ACCEPT [9:952]
-A INPUT -p tcp -m tcp --dport 3888 -j ACCEPT
-A INPUT -p tcp -m tcp --dport 2888 -j ACCEPT
-A INPUT -p tcp -m tcp --dport 2181 -j ACCEPT
COMMIT
# Completed on Sun Mar 27 04:27:48 2022
```

启动Zookeeper集群，三个Zookeeper都要启动，代码如下所示：

```
#在三台服务器中启动Zookeeper服务
[root@server1 ~]# /usr/local/zookeeper/bin/zkServer.sh start
[root@server2 ~]# /usr/local/zookeeper/bin/zkServer.sh start
[root@server3 ~]# /usr/local/zookeeper/bin/zkServer.sh start
#查看集群状态
```

```
[root@server1 ~]# /usr/local/zookeeper/bin/zkServer.sh status
ZooKeeper JMX enabled by default
Using config: /usr/local/zookeeper/bin/../conf/zoo.cfg
Client port found: 2181. Client address: localhost. Client SSL: false.
Mode: follower
[root@server2 ~]# /usr/local/zookeeper/bin/zkServer.sh status
ZooKeeper JMX enabled by default
Using config: /usr/local/zookeeper/bin/../conf/zoo.cfg
Client port found: 2181. Client address: localhost. Client SSL: false.
Mode: leader
[root@server3 data]# /usr/local/zookeeper/bin/zkServer.sh status
ZooKeeper JMX enabled by default
Using config: /usr/local/zookeeper/bin/../conf/zoo.cfg
Client port found: 2181. Client address: localhost. Client SSL: false.
Mode: follower
```

6.2.4 Kafka安装

微课：v6-3 运行Zookeeper集群

1.软件包安装

Kafka通过官方二进制包进行安装，分别在三台服务器中安装。首先安装server1，另外两台使用scp复制软件包的方式完成。安装命令如下：

```
[root@server1 ~]# tar -zxvf kafka_2.12-3.0.1.tgz
kafka_2.12-3.0.1/libs/kafka-streams-3.0.1.jar
kafka_2.12-3.0.1/libs/rocksdbjni-6.19.3.jar
kafka_2.12-3.0.1/libs/kafka-streams-scala_2.12-3.0.1.jar
kafka_2.12-3.0.1/libs/kafka-streams-test-utils-3.0.1.jar
kafka_2.12-3.0.1/libs/kafka-streams-examples-3.0.1.jar
[root@server1 ~]# mv kafka_2.12-3.0.1 /usr/local/kafka
```

2.在kafka目录下创建logs文件夹

在kafka目录下创建logs文件夹，新建命令如下所示：

```
[root@server1 ~]# mkdir /usr/local/kafka/logs
```

3.修改配置文件

修改配置文件的代码如下：

```
[root@server1 ~]# vi /usr/local/kafka/config/server.properties
#broker的全局唯一编号，不能重复
broker.id=0
#删除topic
delete.topic.enable=true
#处理网络请求的线程数量
num.network.threads=3
#处理磁盘I/O的现成数量
num.io.threads=8
#发送套接字的缓冲区大小
```

```
socket.send.buffer.bytes=102400
#接收套接字的缓冲区大小
socket.receive.buffer.bytes=102400
#请求套接字的缓冲区大小
socket.request.max.bytes=104857600
#kafka运行日志存放的路径
log.dirs=/usr/local/kafka/logs
#topic在当前broker上的分区个数
num.partitions=1
#用来恢复和清理data下数据的线程数量
num.recovery.threads.per.data.dir=1
#segment文件保留的最长时间，超时将被删除
log.retention.hours=168
#配置连接Zookeeper集群地址
zookeeper.connect=192.168.13.100:2181,192.168.13.110:2181,192.168.13.
 120:2181/kafka
```

4. 配置环境变量

配置环境变量的代码如下：

```
[root@server1 ~]# vi /etc/profile
#========kafkaconfig===============
export KAFKA_HOME=/usr/local/kafka/
export PATH=$PATH:$KAFKA_HOME/bin
[root@server1 ~]# source /etc/profile
```

5. scp复制安装软件包

scp复制安装软件包的代码如下：

```
[root@server1 ~] # scp -r /usr/local/kafka root@192.168.13.110:/usr/local/
[root@server1 ~] # scp -r /usr/local/kafka root@192.168.13.120:/usr/local/
[root@server1 ~] # scp -r /etc/profile root@192.168.13.110:/etc/profile
[root@server1 ~] # scp -r /etc/profile root@192.168.13.120:/etc/profile
```

6. 分别在server2和server3上修改配置文件

分别在server2和server3上修改配置文件的代码如下：

```
#broker.id不得重复
[root@server2 ~]# vi /usr/local/kafka/config/server.properties
# The id of the broker. This must be set to a unique integer for each
broker.
broker.id=1
[root@server3 ~]# vi /usr/local/kafka/config/server.properties
# The id of the broker. This must be set to a unique integer for each
broker.
broker.id=2
```

7.启动集群

依次在server1、server2、server3节点上启动Kafka，代码如下所示：

```
[root@server1 ~]# source /etc/profile
[root@server1 ~]# kafka-server-start.sh -daemon $KAFKA_HOME/config/
server.properties
[root@server2 ~]# source /etc/profile
[root@server2 ~]# kafka-server-start.sh -daemon $KAFKA_HOME/config/
server.properties
[root@server3 ~]# source /etc/profile
[root@server3 ~]# kafka-server-start.sh -daemon $KAFKA_HOME/config/
server.properties
```

8. 关闭集群

关闭集群的代码如下：

```
[root@server1 ~]# kafka-server-stop.sh
[root@server2 ~]# kafka-server-stop.sh
[root@server3 ~]# kafka-server-stop.sh
```

6.2.5 Kafka命令行操作

1. 查看当前服务器中的所有topic

查看当前服务器中的所有topic的代码如下：

```
[root@server1 ~]# bin/kafka-topics.sh --zookeeper server1:2181/kafka
--list
```

2. 创建topic

创建topic的代码如下：

```
[root@server1 ~]#bin/kafka-topics.sh --zookeeperserver1:2181/kafka\
--create --replication-factor 3 --partitions 1 --topic first
```

3. 删除topic

需要在server.properties中设置delete.topic.enable=true，否则只是标记删除，代码如下：

```
[root@server1 ~]#bin/kafka-topics.sh --zookeeper server1:2181/kafka \
--delete --topic first
```

4. 发送消息

发送消息的代码如下：

```
[root@server1 ~]#bin/kafka-console-producer.sh \
--broker-list server1:9092 --topic first
>hello world
>cqsx  cqsx
```

5. 消费消息

消费消息的代码如下：

```
[root@server1 ~]# bin/kafka-console-consumer.sh\
--bootstrap-server server1:9092 --from-beginning --topic first
```

6. 查看某个topic的详情

查看某个topic的详情的代码如下：

```
[root@server1 ~]#bin/kafka-topics.sh --zookeeper server1:2181/kafka\
--describe --topic first
```

7. 修改分区数

修改分区数的代码如下：

```
[root@server1 ~]#kafka-topics.sh --zookeeper server1:2181/kafka --alter
--topic first --partitions 6
```

6.2.6 Consumer基础配置

每个Consumer进程都会划归到一个逻辑的Consumer Group中，逻辑的订阅者是Consumer Group。当一条消息可以被多个订阅该消息所在的topic的每一个Consumer Group所消费，就好像这条消息被广播到每个Consumer Group一样。在每个Consumer Group中，类似于Queue的概念，即一条消息只会被Consumer Group中的一个Consumer消费，核心配置文件是group.id zookeeper.connect，代码如下所示：

```
## Consumer归属的组ID，broker根据group.id来判断是队列模式还是发布订阅模式
group.id
## 消费者的ID，若没有设置，则会自增
consumer.id
## 一个用于跟踪调查的ID ，最好与group.id相同
client.id = group id value
## 对于Zookeeper集群的指定，可以是多个 hostname1:port1,hostname2:port2,
hostname3:port3
必须与broker使用同样的zk配置
 zookeeper.connect=localhost:2182
## Zookeeper的心跳超时时间，超过这个时间就认为是dead消费者
 zookeeper.session.timeout.ms =6000
## Zookeeper的等待连接时间
 zookeeper.connection.timeout.ms =6000
## Zookeeper的follower与leader的同步时间
 zookeeper.sync.time.ms =2000
## 当Zookeeper中没有初始offset时的处理方式。smallest：重置为最小值；
largest:重置为最大值；anythingelse：抛出异常
 auto.offset.reset = largest
## socket的超时时间，实际的超时时间是：max.fetch.wait + socket.timeout.ms.
```

```
socket.timeout.ms=30*1000
## socket接受的缓存空间大小
socket.receive.buffer.bytes=64*1024
##受从每个分区获取的消息大小限制
fetch.message.max.bytes =1024*1024
## 是否在消费消息后将offset同步到Zookeeper，当Consumer失败后，就能从Zookeeper获
## 取最新的offset
auto.commit.enable =true
## 自动提交的时间间隔
auto.commit.interval.ms =60*1000
## 用来处理消费消息的块，每个块等同于fetch.message.max.bytes中的数值
queued.max.message.chunks =10
## 当有新的Consumer加入Group时，将会reblance，此后将会有partitions的消费端迁移
## 到新的Consumer上，如果一个Consumer获得了某个partition的消费权限，那么它会向zk
## 注册"Partition Owner registry"节点信息，但是有可能此时旧的Consumer尚没有释
## 放此节点
## 此值用于控制注册节点的重试次数
rebalance.max.retries =4
## 每次平衡的时间间隔
rebalance.backoff.ms =2000
## 每次重新选举leader的时间
refresh.leader.backoff.ms
## server发送到消费端的最小数据，若不满足这个数值，则会等待，直到满足要求
fetch.min.bytes =1
## 若不满足最小大小(fetch.min.bytes)，则等待消费端请求的最长等待时间
fetch.wait.max.ms =100
## 指定时间内没有消息到达就抛出异常，一般不需要更改
consumer.timeout.ms = -1
```

6.2.7 Producer基础配置

配置过程中分优先级高、中、低几个级别，一组host和port用于初始化连接。不管这里配置了多少台服务器，都只是用作发现整个集群的全部服务器信息。配置不需要包含集群所有的机器信息，基础配置代码如下所示：

```
## 核心配置：metadata.broker.list request.required.acks producer.type
   serializer.class
## 消费者获取消息元信息(topics, partitions and replicas)的地址，配置格式
是：host1:port1,host2:port2，也可以在外面设置一个vip
metadata.broker.list
## 消息的确认模式
## 0：不保证消息的到达时间，只管发送，低延迟会出现消息的丢失，在某个服务器失败的
## 情况下，有点像TCP
## 1：发送消息，并会等待leader收到消息确认，一定的可靠性
## -1：发送消息，等待leader收到消息确认后再进行复制操作，返回，最高的可靠性
request.required.acks =0
## 消息发送的最长等待时间
request.timeout.ms =10000
## socket的缓存空间大小
```

```
send.buffer.bytes=100*1024
## key的序列化方式，若没有设置，则同serializer.class key.serializer.class
## 分区策略，默认为取模
partitioner.class=kafka.producer.DefaultPartitioner
## 消息的压缩模式，默认为none，可以有gzip和snappy
compression.codec = none
## 可以针对特定的topic进行压缩
compressed.topics=null
## 消息发送失败后的重试次数
message.send.max.retries =3
## 每次失败后的时间间隔
retry.backoff.ms =100
## 生产者定时更新topic元信息的时间间隔，若设置为0，那么会在每个消息发送后都去更新数据
topic.metadata.refresh.interval.ms =600*1000
## 用户随意指定，但是不能重复，主要用于跟踪记录消息
client.id=””
## 生产者的类型：async表示异步执行消息的发送，sync表示同步执行消息的发送
 producer.type=sync
## 异步模式下，会在设置的时间缓存消息，并一次性发送
queue.buffering.max.ms =5000
## 异步模式下，等待时间最长的消息数
queue.buffering.max.messages =10000
## 异步模式下，进入队列的等待时间，若设置为0，要么进入队列，要么直接抛弃
 queue.enqueue.timeout.ms = -1
## 异步模式下，每次发送的最大消息数，前提是触发了queue.buffering.max.messages
## 或受queue.buffering.max.ms的限制
 batch.num.messages=200
## 消息体的系列化处理类 ，转化为字节流进行传输
 serializer.class= kafka.serializer.DefaultEncoder
```

6.3 Kafka API简介

Producer API允许应用程序发布一串流式的数据到一个或者多个Kafka Topic中。Consumer API允许应用程序订阅一个或多个Topic，并且对发布给它们的流式数据进行处理。

Streams API允许应用程序作为流处理器消费一个或者多个Topic产生的输入流，然后产生一个输出流到一个或多个Topic中去，在输入/输出流中进行有效转换。

Connector API允许构建并运行可重用的生产者或者消费者，将Kafka Topics连接到已存在的应用程序或者数据系统中。比如，连接一个关系型数据库，捕捉表的所有变更内容。

在Kafka中，客户端和服务器之间的通信是通过简单、高性能、与语言无关的TCP协议完成的。此协议已版本化并保持与旧版本的向后兼容性。Kafka提供多种语言的客户端。客户端架构如图6-1所示。

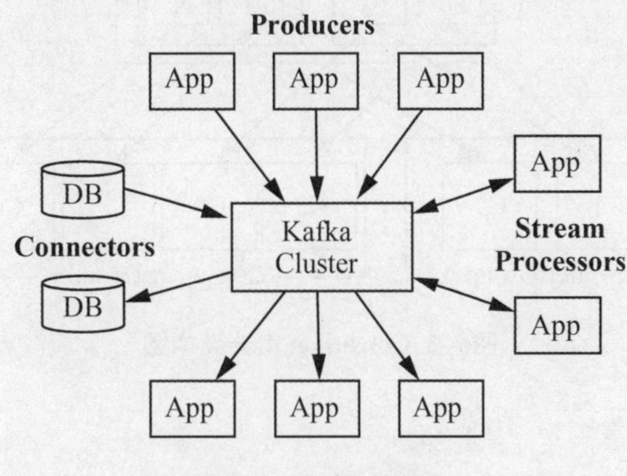

图6-1 客户端架构

6.3.1 Kafka API Producer

Producer会为每个Partition维护一个缓冲，用来记录还没有发送的数据。缓冲区的大小用batch.size指定，数据通过Buffer传至Partition中。Producer数据流程图如图6-2所示。

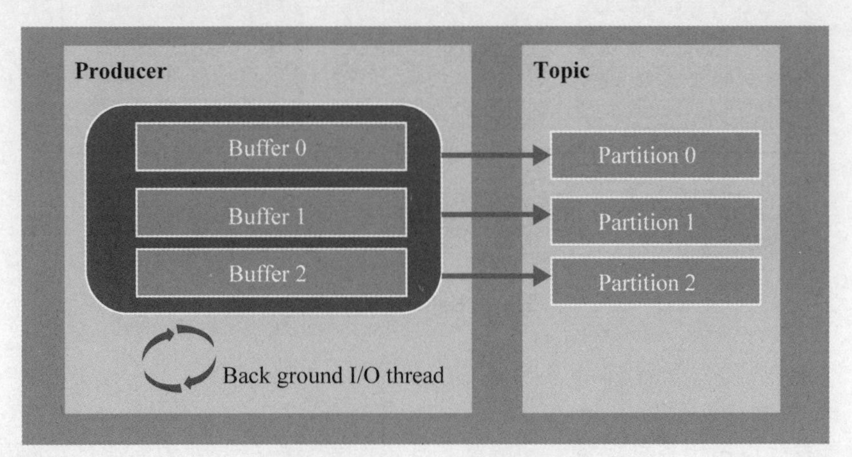

图6-2 Producer数据流程图

在生产开发环境中，Kafka API的默认值为16 KB，linger.ms为响应配置参数，Buffer中的数据在达到batch.size前需要等待一段时间，acks为配置请求成功的标准。

6.3.2 Kafka API Consumer

简单的Consumer位于kafka.javaapi.consumer包中，不提供负载均衡容错的特性。每次获取数据都要指定Topic、Partition、Offset、Fetch的大小。高级别的Consumer客户端通过与Broker交互来实现Consumer Group的负载均衡。Consumer集群架构图如图

6-3所示。

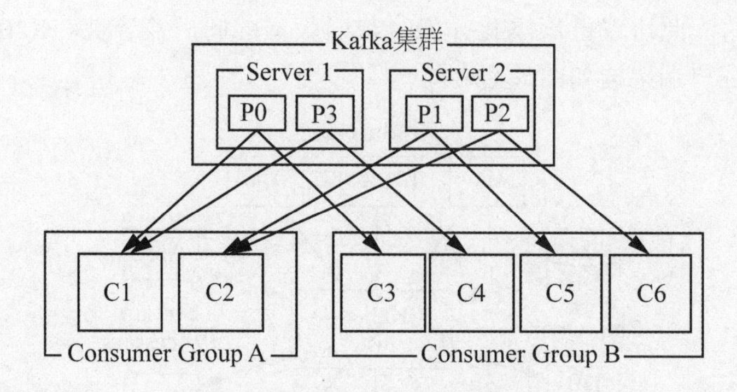

图6-3 Consumer集群架构图

6.3.3 体系架构

在系统开发过程中，程序的编写应注重服务高可用性、调用可靠性等，需对整体架构进行高可用性规划。在实际设计过程中，与Hadoop集群建立通信，采用分布式集群架构，有高效数据信息流的吞吐量。Kafka体系架构如图6-4所示。

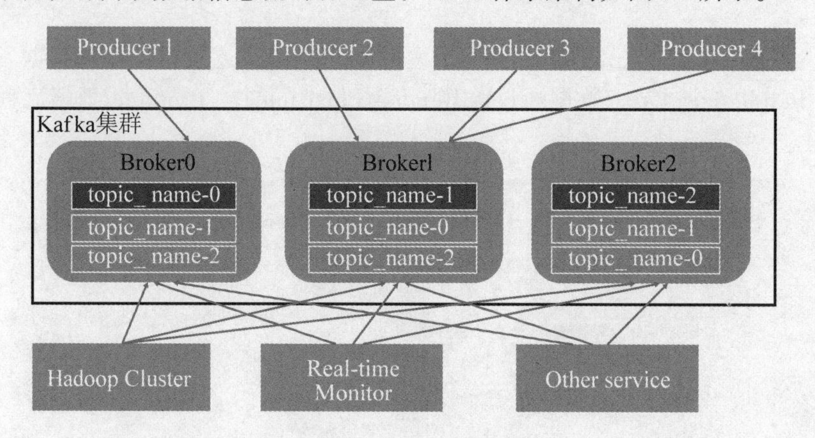

图6-4 Kafka体系架构

6.3.4 Kafka技术实现

消息系统可用于各种场景，如解耦数据生产者、缓存未处理的消息。Kafka可作为传统消息系统的替代者，与传统消息系统相比，Kafka有更好的吞吐量和更好的可用性，这有利于处理大规模的消息。根据经验，消息传递对吞吐量要求较低，可能要求较低的端到端延迟，并经常依赖Kafka的可靠Durable机制。因此，Kafka可以与传统的消息系统ActiveMQ和RabbitMQ相媲美。

写入Kafka中的数据存入了磁盘，并且有冗余备份，Kafka允许Producer等待确认，通过配置，只有等到所有的replication完成复制才算写入成功，这样可保证数据的可用性。Kafka认真对待存储，并允许客户端自行控制读取位置。你可以认为Kafka是一个特殊的文件系统，它能够提供高性能、低延迟、高可用的日志提交功能。

日志系统一般具备日志的收集、清洗、聚合、存储展示等功能。Kafka常用来替代其他日志聚合解决方案。与Scribe Flume相比，Kafka可提供更好的性能、更低的端到端延迟。与目前流行的日志系统ELK比较，Kafka与ELK的配合才是更成熟的方案。Kafka在ELK技术栈中，主要起到Buffer的作用，必要时可进行日志的汇流。

Kafka的最初作用是将用户行为跟踪管道重构为一组实时发布-订阅源，将网站活动（浏览网页、搜索或其他用户操作）发布到中心Topics中，每种活动类型对应一个Topic。基于这些订阅源，能够实现一系列应用，如批量将Kafka的数据加载到Hadoop或离线数据仓库系统，进行离线数据处理并生成报告。用户浏览网页时，会生成许多活动信息，因此，活动跟踪的数据量通常非常大。

6.4 Kafka监控

Kafka Eagle监控系统是一款用来监控Kafka集群的工具。目前最新的版本是v1.2.3，支持管理多个Kafka集群。Kafka主题包含查看、删除、创建等消费者组合、消费者实例监控、消息阻塞告警、Kafka集群健康状态查看等。目前，Kafka Eagle v1.2.3版本整合了前面介绍的所需功能。

6.4.1 Kafka Eagle版本介绍

在安装Kafka Eagle监控系统之前，需要准备好Kafka Eagle安装包。这里有两种方式的软件包，即已编译好的安装包、源代码自行编译安装包。官方下载链接为http://download.kafka-eagle.org/。

6.4.2 Kafka Eagle安装

微课：
v6-4 监控
Kafka Eagle
安装

1. 基础环境说明

在server1中，安装环境需要Java的支持，默认Kafka的安装环境已安装JDK。

2. 安装命令

安装命令如下所示：

```
[root@server1 ~]# tar -zxvf kafka-eagle-bin-2.1.0.tar.gz
kafka-eagle-bin-2.1.0/
```

```
kafka-eagle-bin-2.1.0/efak-web-2.1.0-bin.tar.gz
[root@server1 ~]# mv kafka-eagle-bin-2.1.0 /usr/local/kafka-eagle
[root@server1 kafka-eagle]# tar -xvf efak-web-2.1.0-bin.tar.gz
efak-web-2.1.0/kms/conf/jaspic-providers.xsd
efak-web-2.1.0/kms/conf/context.xml
efak-web-2.1.0/kms/conf/catalina.policy
efak-web-2.1.0/kms/conf/web.xml
efak-web-2.1.0/kms/conf/tomcat-users.xsd
efak-web-2.1.0/kms/webapps/ke.war
#修改环境变量
[root@server1 ~]# vi /etc/profile
#=======efak================
export KE_HOME=/usr/local/kafka-eagle/efak-web-2.1.0
export PATH=$PATH:$KE_HOME/bin
```

3. 修改配置文件

根据Kafka集群的实际情况配置EFAK，例如，Zookeeper地址、Kafka集群的版本类型（zk为低版本、Kafka为高版本）、开启安全认证的Kafka集群等。

```
[root@server1 ~]# vi /usr/local/kafka-eagle/efak-web-2.1.0/conf/
system-config.properties
######################################
# multi zookeeper & kafka cluster list
# Settings prefixed with 'kafka.eagle.' will be deprecated, use
'efak.' instead
######################################
efak.zk.cluster.alias=cluster1,cluster2
cluster1.zk.list=192.168.13.100:2181,192.168.13.110:2181,192.168.13.
120:2181
cluster2.zk.list=xdn10:2181,xdn11:2181,xdn12:2181
######################################
# zookeeper enable acl
"/usr/local/kafka-eagle/efak-web-2.1.0/conf/system-config.properties"
128L, 4272C
######################################
# multi zookeeper & kafka cluster list
# Settings prefixed with 'kafka.eagle.' will be deprecated, use
'efak.' instead
######################################
efak.zk.cluster.alias=cluster1,cluster2
cluster1.zk.list=tdn1:2181,tdn2:2181,tdn3:2181
cluster2.zk.list=xdn10:2181,xdn11:2181,xdn12:2181
######################################
# zookeeper enable acl
######################################
cluster1.zk.acl.enable=false
cluster1.zk.acl.schema=digest
cluster1.zk.acl.username=test
cluster1.zk.acl.password=test123
######################################
# broker size online list
```

```
######################################
cluster1.efak.broker.size=20
######################################
# zk client thread limit
######################################
kafka.zk.limit.size=16
######################################
# EFAK webui port
######################################
efak.webui.port=8048
######################################
# EFAK enable distributed
######################################
efak.distributed.enable=false
efak.cluster.mode.status=master
efak.worknode.master.host=localhost
efak.worknode.port=8085
######################################
# kafka jmx acl and ssl authenticate
######################################
cluster1.efak.jmx.acl=false
cluster1.efak.jmx.user=keadmin
cluster1.efak.jmx.password=keadmin123
cluster1.efak.jmx.ssl=false
cluster1.efak.jmx.truststore.location=/data/ssl/certificates/kafka.
truststore
cluster1.efak.jmx.truststore.password=ke123456
######################################
# kafka offset storage
######################################
cluster1.efak.offset.storage=kafka
cluster2.efak.offset.storage=zk
######################################
# kafka jmx uri
######################################
# kafka metrics, 15 days by default
######################################
efak.metrics.charts=true
efak.metrics.retain=15
######################################
efak.sql.topic.records.max=5000
efak.sql.topic.preview.records.max=10
######################################
# delete kafka topic token
######################################
efak.topic.token=keadmin
######################################
# kafka sasl authenticate
######################################
cluster1.efak.sasl.enable=false
cluster1.efak.sasl.protocol=SASL_PLAINTEXT
cluster1.efak.sasl.mechanism=SCRAM-SHA-256
```

```
cluster1.efak.sasl.client.id=
cluster1.efak.blacklist.topics=
cluster1.efak.sasl.cgroup.enable=false
cluster1.efak.sasl.cgroup.topics=
cluster2.efak.sasl.enable=false
cluster2.efak.sasl.protocol=SASL_PLAINTEXT
cluster2.efak.sasl.mechanism=PLAIN
cluster2.efak.sasl.client.id=
cluster2.efak.blacklist.topics=
cluster2.efak.sasl.cgroup.enable=false
cluster2.efak.sasl.cgroup.topics=
######################################
cluster3.efak.ssl.enable=false
cluster3.efak.ssl.protocol=SSL
cluster3.efak.ssl.truststore.location=
cluster3.efak.ssl.truststore.password=
cluster3.efak.ssl.keystore.location=
cluster3.efak.ssl.keystore.password=
cluster3.efak.ssl.key.password=
cluster3.efak.ssl.endpoint.identification.algorithm=https
cluster3.efak.blacklist.topics=
cluster3.efak.ssl.cgroup.enable=false
cluster3.efak.ssl.cgroup.topics=
######################################
# kafka sqlite jdbc driver address
######################################
#efak.driver=org.sqlite.JDBC
#efak.url=jdbc:sqlite:/hadoop/kafka-eagle/db/ke.db
#efak.username=root
#efak.password=www.kafka-eagle.org
######################################
# kafka mysql jdbc driver address
######################################
efak.driver=com.mysql.cj.jdbc.Driver
efak.url=jdbc:mysql://127.0.0.1:3306/ke?useUnicode=true&characterEncoding=
UTF-8&zeroDateTimeBeha
vior=convertToNull
efak.username=root
efak.password=123456
```

4. 启动服务

启动服务的执行命令如下所示：

```
[root@server1 efak-web-2.1.0]# cd bin/
[root@server1 bin]# ll
total 24
-rw-r--r--. 1 root root  1848 Sep 12  2021 ke.bat
-rwxr-xr-x. 1 root root 11289 Dec 12 09:08 ke.sh
-rwxr-xr-x. 1 root root  4776 Dec 12 09:28 worknode.sh
[root@server1 bin]# source /etc/profile
[root@server1 bin]# chmod 777 ke.sh
[root@server1 ~]# ke.sh start
[2022-03-27 09:55:01] INFO: Startup Progress: [############
#################################### [2022-03-27 09:55:01]
INFO: Startup Progress: [################################
```

```
################ [2022-03-27 09:55:01] INFO: Startup
Progress: [##################################################[2022-
03-27 09:55:01] INFO: Startup Progress:
[##################################################[2022-03-27 09:55:01]
INFO: Startup Progress: [##############################################
####] | 100%
[2022-03-27 09:54:39] INFO: Status Code[0]
[2022-03-27 09:54:39] INFO: [Job done!]
Welcome to
    ______     ______       ___     __ __
   / ____/    / ____/      /   |   / //_/
  / __/      / /_         / /| |  / ,<
 / /___     / __/        / ___ | / /| |
/_____/    /_/          /_/  |_|/_/ |_|
( Eagle For Apache Kafka® )

Version 2.1.0 -- Copyright 2016-2022
*******************************************************************
* EFAK Service has started success.
* Welcome,Now you can visit 'http://192.168.13.100:8048'
* Account:admin ,Password:123456
*******************************************************************
* <Usage> ke.sh [start|status|stop|restart|stats] </Usage>
* <Usage> https://www.kafka-eagle.org/ </Usage>
*******************************************************************
```

6.4.3 Kafka Eagle访问

微课：v6-5 ke. sh启动过程介绍

通过以上配置步骤，服务器正常运行后，可以通过浏览器访问Eagle监控系统了。访问地址根据环境服务器的IP地址而定，本实训环境访问链接为http://192.168.13.100:8048，访问用户名为admin，密码为123456。Eagle Web UI登录界面如图6-5所示。Eagle Web UI功能界面如图6-6所示。

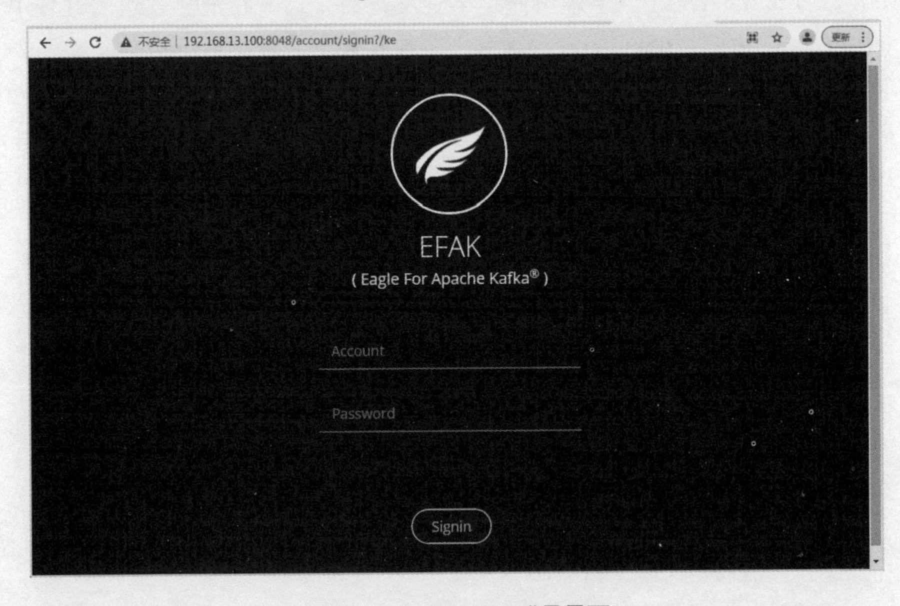

图6-5 Eagle Web UI登录界面

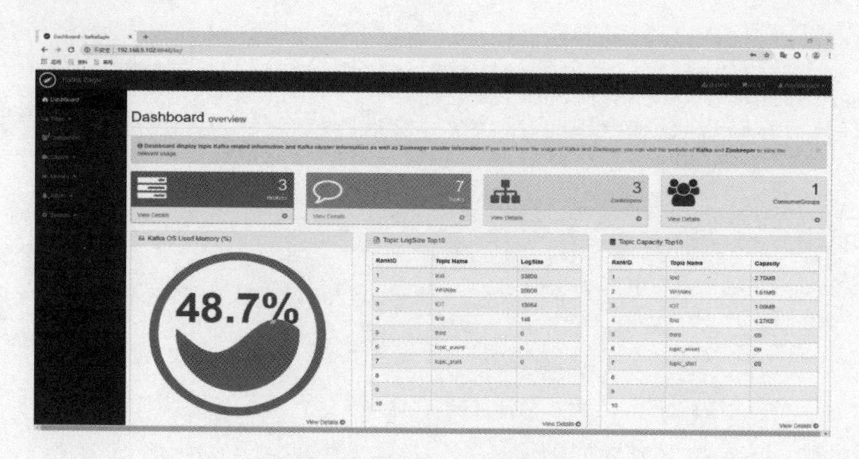

图6-6 Eagle Web UI功能界面

6.5 Kafka编程

Kafka是一个高吞吐量的分布式发布订阅消息系统，可以处理消费者在网站中的所有动作流数据，以容错持久的方式存储记录流。当发生记录流的时候，通过Java编程方式处理程序。

6.5.1 Kafka消息发送流程

Kafka的Producer发送消息采用的是异步发送的方式。在消息发送的过程中，涉及两个线程（Main线程和Sender线程）和一个线程共享变量（RecordAccumulator）。Main线程将消息发送给RecordAccumulator，Sender线程不断地从RecordAccumulator中获取消息并发送到Kafka Broker。需要创建一个生产者对象来发送数据，ProducerConfig获取一系列配置参数ProducerRecord：每个数据都要封装成ProducerRecord对象。

1. 不带回调函数的API

不带回调函数的API实现代码如下：

```
package com.cqsx.kafka;
import org.apache.kafka.clients.producer.*;
import java.util.Properties;
import java.util.concurrent.ExecutionException;
public class CustomProducer {
    public static void main(String[] args) throws ExecutionException,
        InterruptedException {
        Properties props = new Properties();
```

```
        props.put("bootstrap.servers","server1:9092");
        //kafka集群，broker-list
        props.put("acks","all");
        props.put("retries",1);          //重试次数
        props.put("batch.size",16384); //批次大小
        props.put("linger.ms",1);        //等待时间
        props.put("buffer.memory",33554432);//RecordAccumulator缓冲区大小
        props.put("key.serializer",
            "org.apache.kafka.common.serialization.StringSerializer");
        props.put("value.serializer",
            "org.apache.kafka.common.serialization.StringSerializer");
        Producer<String,String> producer = new KafkaProducer<>(props);
        for (int i = 0;i < 100;i++) {
            producer.send(new ProducerRecord<String,String>("first",
                Integer.toString(i),Integer.toString(i)));
        }
        producer.close();
    }
}
```

2.带回调函数的API

回调函数会在Producer收到ack时调用，为异步调用。回调函数有两个参数，分别是RecordMetadata和Exception。如果Exception为null，则说明消息发送成功，如果Exception不为null，则说明消息发送失败。消息发送失败会自动重试，不需要我们在回调函数中手动重试。实现代码如下所示：

```
package com.cqsx.kafka;
import org.apache.kafka.clients.producer.*;
import java.util.Properties;
import java.util.concurrent.ExecutionException;
public class CustomProducer {
    public static void main(String[] args) throws ExecutionException,
        InterruptedException {
        Properties props = new Properties();
        props.put("bootstrap.servers","server1:9092");
        //kafka集群，broker-list
        props.put("acks","all");
        props.put("retries",1);          //重试次数
        props.put("batch.size",16384); //批次大小
        props.put("linger.ms",1);        //等待时间
        props.put("buffer.memory",33554432);//RecordAccumulator缓冲区大小
        props.put("key.serializer",
            "org.apache.kafka.common.serialization.StringSerializer");
        props.put("value.serializer",
            "org.apache.kafka.common.serialization.StringSerializer");
        Producer<String,String> producer = new KafkaProducer<>(props);
        for (int i = 0;i < 100;i++) {
            producer.send(new ProducerRecord<String,String>("first",
                Integer.toString(i),Integer.toString(i)),new Callback() {
                //回调函数会在Producer收到ack时调用，为异步调用
```

```
                @Override
                public void onCompletion(RecordMetadata metadata,
                    Exception exception) {
                    if (exception == null) {
                        System.out.println("success->" +
                            metadata.offset());
                    } else {
                        exception.printStackTrace();
                    }
                }
            });
        }
        producer.close();
    }
}
```

6.5.2 Kafka同步发送API

同步发送的编程思想是，当一条消息发送之后，会阻塞当前线程，直至返回ack。由于send()方法返回的是一个Future对象，根据Future对象的特点，也可以实现同步发送的效果，只需调用Future对象的get()方法即可。实现代码如下所示：

```
package com.cqsx.kafka;
import org.apache.kafka.clients.producer.KafkaProducer;
import org.apache.kafka.clients.producer.Producer;
import org.apache.kafka.clients.producer.ProducerRecord;
import java.util.Properties;
import java.util.concurrent.ExecutionException;
public class CustomProducer {
public static void main(String[] args) throws ExecutionException,
InterruptedException {
        Properties props = new Properties();
        props.put("bootstrap.servers","server1:9092");
        //kafka集群，broker-list
        props.put("acks","all");
        props.put("retries",1);          //重试次数
        props.put("batch.size",16384); //批次大小
        props.put("linger.ms",1);        //等待时间
        props.put("buffer.memory",33554432);//RecordAccumulator缓冲区大小
        props.put("key.serializer",
            "org.apache.kafka.common.serialization.StringSerializer");
        props.put("value.serializer",
            "org.apache.kafka.common.serialization.StringSerializer");
        Producer<String,String> producer = new KafkaProducer<>(props);
        for (int i = 0;i < 100;i++) {
            producer.send(new ProducerRecord<String,String>("first",
                Integer.toString(i),Integer.toString(i))).get();
        }
        producer.close();
    }
}
```

6.5.3 Kafka Consumer

Consumer在消费数据时，可靠性很容易保证，因为数据在Kafka中是持久化的，故不用担心数据丢失问题。由于Consumer在消费过程中可能会出现断电、宕机等故障，在恢复后，需要从故障前的位置继续消费，所以Consumer需要实时记录自己消费到哪个offset，以便故障恢复后继续消费。维护offset是Consumer消费数据时必须考虑的问题。实现代码如下所示：

```
package com.cqsx.kafka;
import org.apache.kafka.clients.consumer.ConsumerRecord;
import org.apache.kafka.clients.consumer.ConsumerRecords;
import org.apache.kafka.clients.consumer.KafkaConsumer;
import java.util.Arrays;
import java.util.Properties;
public class CustomConsumer {
    public static void main(String[] args) {
        Properties props = new Properties();
        props.put("bootstrap.servers","server1:9092");
        props.put("group.id","test");
        props.put("enable.auto.commit","true");
        props.put("auto.commit.interval.ms","1000");
        props.put("key.deserializer",
            "org.apache.kafka.common.serialization.StringDeserializer");
        props.put("value.deserializer",
            "org.apache.kafka.common.serialization.StringDeserializer");
        KafkaConsumer<String,String> consumer = new KafkaConsumer<>(props);
        consumer.subscribe(Arrays.asList("first"));
        while (true) {
            ConsumerRecords<String,String> records = consumer.poll(100);
            for (ConsumerRecord<String, String> record:records)
                System.out.printf("offset = %d,key = %s,value = %s%n",
                    record.offset(),record.key(),record.value());
        }
    }
}
```

6.5.4 Kafka手动提交offset

虽然自动提交offset十分便利，但是，由于其是基于时间提交的，开发人员难以把握offset提交的时机，因此Kafka还提供了手动提交offset的API。手动提交offset的方法有两种：commitSync（同步提交）和commitAsync（异步提交）。两者的相同点是都会提交本次poll的一批数据最高的偏移量；两者的不同点是，commitSync阻塞当前线程，直到提交成功，并且有失败重试机制，当出现不可控因素时，也会出现提交失败；而commitAsync则没有失败重试机制，故有可能提交失败。实现代码分同步提交offset和异步提交offset。

1. 同步提交offset

由于同步提交offset有失败重试机制，因此更加可靠。以下为同步提交offset的代码：

```
package com.cqsx.kafka.consumer;
import org.apache.kafka.clients.consumer.ConsumerRecord;
import org.apache.kafka.clients.consumer.ConsumerRecords;
import org.apache.kafka.clients.consumer.KafkaConsumer;
import java.util.Arrays;
import java.util.Properties;
/**
 * @author liubo
 */
public class CustomComsumer {
    public static void main(String[] args) {
        Properties props = new Properties();
        props.put("bootstrap.servers","server1:9092");//Kafka集群
        props.put("group.id","test");
        //消费者组，只要group.id相同，就属于同一个消费者组
        props.put("enable.auto.commit","false");//关闭自动提交offset
        props.put("key.deserializer",
            "org.apache.kafka.common.serialization.StringDeserializer");
        props.put("value.deserializer",
            "org.apache.kafka.common.serialization.StringDeserializer");
        KafkaConsumer<String,String> consumer = new KafkaConsumer<>(props);
        consumer.subscribe(Arrays.asList("first"));
        //消费者订阅主题
        while (true) {
            ConsumerRecords<String,String> records = consumer.poll(100);
            //消费者拉取数据
            for (ConsumerRecord<String,String> record:records) {
                System.out.printf("offset = %d,key = %s,value = %s%n",
                    record.offset(),record.key(),record.value());
            }
            consumer.commitSync();
            //同步提交，当前线程会阻塞直到offset提交成功
        }
    }
}
```

2. 异步提交offset

虽然同步提交offset更可靠，但是会阻塞当前线程，直到提交成功。因此，吞吐量会受到很大的影响。大多数情况下，会选用异步提交offset的方式。以下为异步提交offset的代码：

```
package com.cqsx.kafka.consumer;
import org.apache.kafka.clients.consumer.*;
import org.apache.kafka.common.TopicPartition;
import java.util.Arrays;
import java.util.Map;
import java.util.Properties;
/**
 * @author liubo
 */
public class CustomConsumer {
```

```
    public static void main(String[] args) {
        Properties props = new Properties();
        props.put("bootstrap.servers","server1:9092");//Kafka集群
        props.put("group.id","test");
        //消费者组，只要group.id相同，就属于同一个消费者组
        props.put("enable.auto.commit","false");//关闭自动提交offset
        props.put("key.deserializer",
            "org.apache.kafka.common.serialization.StringDeserializer");
        props.put("value.deserializer",
            "org.apache.kafka.common.serialization.StringDeserializer");
        KafkaConsumer<String,String> consumer = new KafkaConsumer<>(props);
        consumer.subscribe(Arrays.asList("first"));//消费者订阅主题
        while (true) {
            ConsumerRecords<String,String> records = consumer.poll(100);
            //消费者拉取数据
            for (ConsumerRecord<String, String> record:records) {
                System.out.printf("offset = %d,key = %s,
                    value = %s%n",record.offset(),record.key(),
                    record.value());
            }
            consumer.commitAsync(new OffsetCommitCallback() {
                @Override
                public void onComplete(Map<TopicPartition,
                    OffsetAndMetadata>
                    offsets,Exception exception) {
                    if (exception != null) {
                        System.err.println("Commit failed for" + offsets);
                    }
                }
            });//异步提交
        }
    }
}
```

6.6 本章小结

本章首先介绍了Kafka开源实时流数据处理平台，使用Java语言编写API。其次介绍了配置Kafka集群高吞吐量的分布式发布订阅消息系统，实现可以处理消费者在网站中的所有动作流数据。最后介绍了Kafka的基础安装配置和应用管理，以及基础的项目案例开发实战。

6.7 课后习题

1. 选择题

（1）Kafka的开发公司是（　　）。

A.Apache　　B.Hadoop　　C.Jaykreps　　D.Baidu

（2）以下选项不是Kafka API的是（　　）。

A.Producer　　B.date　　C.Connector　　D.Streams

（3）Kafka Eagle的默认访问端口是（　　）。

A.80　　B.8080　　C.8048　　D.8081

（4）Kafka Eagle的默认访问用户名是（　　）。

A.root　　B.Admin　　C.guest　　D.admin

（5）kafka-console-producer的默认访问端口是（　　）。

A.9090　　B.8080　　C.9092　　D.900

2. 简答题

（1）简述Kafka的作用及其特点。

（2）简述Kafka常用的编程API及其功能。

（3）简述Kafka Eagle监控系统的功能特点。

第7章 PySpark开发应用

学习目标

（1）通过本章的学习，让初学Spark的读者能够了解实时计算技术。

（2）由浅入深地认识实时计算的概念。主要用Python语言开发Spark实时计算程序。

（3）主要介绍PySpark基础部署，为进一步了解目前主流实时计算打下基础。

思政目标

（1）在网络化、信息化和城市化的推动下，大数据技术知识不断增加，学习也在日益多样化。把握学习机会，努力实现个人社会价值。

（2）实时计算技术时刻都在改变我们的生活，跟随新兴技术的步伐，用知识改变我们身边的每一个人，给他们带来便利。

7.1 PySpark概述

7.1.1 PySpark简介

Apache Spark是使用Scala语言编写的一个计算框架。为了支持Python语言使用Spark，Apache Spark社区开发了一款工具PySpark。利用PySpark中的Py4j库，我们可以通过Python语言操作RDDs。Apache Spark涵盖了数据驱动的基本功能以及讲述了如何使用其各种组件。本书主要针对那些想要从事实时计算框架编程的读者。本书的目的是让读者能够轻松地了解PySpark的基本功能并快速入门。本书中我们假定读者已经有了一些基本的编程语言基础以及了解了什么是编程框架。此外，如果读者有Apache Spark、Hadoop、Scala、HDFS和Python的基础，那么学习下面的内容将会事半功倍。

Apache Spark是一个流行的实时处理框架，它可以通过内存计算的方式来实时进行数据分析。它起源于Apache Hadoop MapReduce，然而，Apache Hadoop MapReduce只能进行批处理，但是无法实现实时计算。为了弥补这一缺陷，Apache Spark对其进行了扩展，除批处理和实时计算外，Apache Spark还支持交互式查询与迭代式算法等特性。此外，Apache Spark由自己的集群管理方式来支持其应用。Apache Spark利用了Apache Hadoop的存储和计算功能，同时，也使用了HDFS来存储并且可以通过YARN来运行Spark应用。

PySpark提供了PySpark Shell，它是一款结合了Python API和Spark Core的工具，同时能够初始化Spark环境。目前，由于Python具有丰富的扩展库，所以大多数数据科学家和数据分析从业人员都在使用Python。

7.1.2 PySpark与生活

使用Spark处理天气数据，通过使用中央气象台官方网站中的数据，主要是最近24小时各个城市的天气数据，包括时间点、整点气温、整点降水量、风力、整点气压、相对湿度等。正常情况下，每个城市会对应24条数据（每个整点一条），主要计算了各个城市过去24小时的平均气温和降水量情况。通过Spark实时处理技术对天气数据进行分析后，就得到了我们关注的天气预报数据。案例程序通过Spark能够很快得到有用的数据，但在开发程序时，会花大量的时间和精力去研究。类似天气预报这样的数据在我们的生活中比比皆是，作为初学者的读者们，应该静下心来学习，很多技术时刻都在改变着我们的生活。

7.2 PySpark配置

PySpark包含在Apache Spark网站上提供的Spark官方版本中。对于Python开发用户，PySpark还提供pip从PyPI安装。这通常用于本地使用或作为客户端连接到集群。

7.2.1 下载Spark

微课：
v7-1 基础环境PySpark安装

本章介绍的软件资源建议从官方网站下载，并下载最新版本。通常官方发布的版本为稳定版本，官方详细的下载链接为https://spark.apache.org/downloads.html，本书测试使用的软件包名为spark-3.2.1-bin-hadoop3.2.tgz。

7.2.2 安装配置

从官方网址中将Apache Spark spark-3.2.1-bin-hadoop3.2.tgz下载到本地服务器

中，解压软件包配置环境变量。代码如下所示：

```
[root@server1 ~]# tar -zxvf spark-3.2.1-bin-hadoop3.2.tgz
spark-3.2.1-bin-hadoop3.2/sbin/start-thriftserver.sh
spark-3.2.1-bin-hadoop3.2/sbin/start-worker.sh
spark-3.2.1-bin-hadoop3.2/sbin/start-workers.sh
spark-3.2.1-bin-hadoop3.2/sbin/stop-all.sh
spark-3.2.1-bin-hadoop3.2/sbin/stop-history-server.sh
spark-3.2.1-bin-hadoop3.2/sbin/stop-master.sh
spark-3.2.1-bin-hadoop3.2/sbin/stop-mesos-dispatcher.sh
spark-3.2.1-bin-hadoop3.2/sbin/stop-mesos-shuffle-service.sh
spark-3.2.1-bin-hadoop3.2/sbin/stop-slave.sh
spark-3.2.1-bin-hadoop3.2/sbin/stop-slaves.sh
spark-3.2.1-bin-hadoop3.2/sbin/stop-thriftserver.sh
spark-3.2.1-bin-hadoop3.2/sbin/stop-worker.sh
spark-3.2.1-bin-hadoop3.2/sbin/stop-workers.sh
spark-3.2.1-bin-hadoop3.2/sbin/workers.sh
spark-3.2.1-bin-hadoop3.2/yarn/
spark-3.2.1-bin-hadoop3.2/yarn/spark-3.2.1-yarn-shuffle.jar
#复制文件到/usr目录中
[root@server1 ~]# mv spark-3.2.1-bin-hadoop3.2 /usr/local/
[root@server1 ~]#vi /etc/profile
#将以下代码添加到文件中
export SPARK_HOME=/usr/local/spark-3.2.1-bin-hadoop3.2/
export PATH=$SPARK_HOME/bin:$PATH
export PYTHONPATH=$SPARK_HOME/python/:$PYTHONPATH
[root@server1 ~]# source /etc/profile
[root@server1 python3]# pyspark
Python 3.6.2 (default,Mar 28 2022,00:25:27)
[GCC 4.8.5 20150623 (Red Hat 4.8.5-44)] on linux
Type "help","copyright","credits" or "license" for more
information.
Using Spark's default log4j profile:org/apache/spark/log4j-defaults.
properties
Setting default log level to "WARN".
To adjust logging level use sc.setLogLevel(newLevel).For SparkR,use
setLogLevel(newLevel).
22/03/28 00:26:54 WARN NativeCodeLoader:Unable to load native-hadoop
library for your platform...using builtin-java classes where applicable
/usr/local/spark-3.2.1-bin-hadoop3.2/python/pyspark/context.py:238:Fu
tureWarning:Python 3.6 support is deprecated in Spark 3.2.
    FutureWarning
Welcome to
      ____              __
     / __/__  ___ _____/ /__
    _\ \/ _ \/ _ `/ __/  '_/
   /__ / .__/\_,_/_/ /_/\_\   version 3.2.1
      /_/

Using Python version 3.6.2 (default,Mar 28 2022 00:25:27)
Spark context Web UI available at http://server1:4040
Spark context available as 'sc' (master = local[*],app id =
```

```
local-1648441619043).
SparkSession available as 'spark'.
```

7.3 PySpark常用接口

微课：v7-2 运行PySpark

PySpark是Python中Apache Spark的接口。它不仅允许使用Python API编写Spark应用程序，还提供PySpark Shell在分布式环境中交互式分析你的数据。PySpark支持Spark的大部分功能，例如Spark SQL、DataFrame、Streaming、MLlib机器学习和Spark Core。

7.3.1 RDD

每个Spark应用程序都包含一个驱动程序，该驱动程序运行用户的功能并在集群上执行各种并行操作。Spark提供的主要抽象是弹性分布式数据集(Resilient Distributed Datasets，RDD)，它是跨集群节点分区的元素集合，可以并行操作。RDD是从Hadoop文件系统或其他Hadoop支持的文件系统或驱动程序中现有的Scala集合开始，并对其进行转换来创建的。用户还可以要求Spark将RDD持久化到内存中，以便在并行操作中有效地重用它。最后，RDD会自动从节点故障中恢复。Spark中的第二个抽象是可以在并行操作中使用的共享变量。默认情况下，当Spark在不同的节点上并行运行一个函数作为一组任务时，它会将函数中使用的每个变量的副本发送到每个任务。有时，需要在任务之间或在任务和驱动程序之间共享变量。Spark支持两种类型的共享变量，即累加器和广播变量，广播变量可在所有节点的内存中缓存。

1. 与Spark链接

Spark 3.2.1适用于Python 3.6+。Spark 3.2.1可以使用标准的CPython解释器，因此可以使用NumPy等C库。它也适用于PyPy 2.3+，Spark 3.1.0中删除了对Python 2、Python 3.4和Python 3.5的支持。Spark 3.2.0中已弃用Python 3.6的支持。Python中的Spark应用程序可以在运行时使用bin/spark-submit包含Spark的脚本，也可以将其包含在setup.py中，代码如下所示：

```
install_requires=[
    'pyspark=={site.SPARK_VERSION}'
]
```

2. 初始化Spark

Spark程序必须做的第一件事是创建一个SparkContext对象，并告诉Spark如何访问集群。要创建一个SparkContext对象，首先要构建一个包含有关应用程序信息的

SparkConf对象，代码如下所示：

```
conf = SparkConf().setAppName(appName).setMaster(master)
sc = SparkContext(conf=conf) "spark",
    "akka",
    "spark vs hadoop",
```

3. 使用Shell

在PySpark Shell中已经创建了一个特殊的解释器用来感知SparkContext，变量名是sc。若创建你自己的SparkContext，它是不会生效的。可以使用master参数设置上下文连接到哪个主服务器，也可以通过将逗号分隔列表传递给py-files，有关第三方Python依赖项请参阅Python包管理，还可以通过向packages参数提供逗号分隔的Maven坐标列表来将依赖项添加到Shell会话中。任何可能存在依赖关系的附加存储库都可以传递给repositories参数。Shell执行命令如下所示：

```
#./bin/pyspark--master local
#./bin/pyspark--master local[4]--py-files code.py
```

可以在增强的Python解释器IPython中启动PySpark Shell。PySpark适用于IPython 1.0.0及更高版本。要使用IPython，请在运行时将PYSPARK_DRIVER_PYTHON变量设置为ipythonbin/pyspark，执行命令如下所示：

```
#PYSPARK_DRIVER_PYTHON=ipython./bin/pyspark
#PYSPARK_DRIVER_PYTHON=jupyter PYSPARK_DRIVER_PYTHON_OPTS=notebook./
bin/pyspark
```

4. 外部数据集

PySpark可以从Hadoop支持的任何存储源创建分布式数据集，包括本地文件系统、HDFS、Cassandra、HBase、Amazon S3等。Spark支持文本文件、SequenceFiles和其他Hadoop InputFormat。可以使用SparkContext的textFile方法创建文本文件RDD。此方法获取文件的URI并将其按行读取为一个集合。以下是一个程序调用命令：

```
>>> distFile = sc.textFile("data.txt")
```

5. 使用键–值对

虽然大多数RDD操作都包含任意类型对象，但少数特殊操作仅在键–值对的RDD上可用。常见的是分布式“shuffle”操作，例如通过键对元素进行分组或聚合。在Python中，这些操作在包含元组的RDD上运行，例如(1,2)。只需创建这样的元组，然后调用你想要的操作。以下代码使用reduceByKey来计算文件中每行文本出现的次数：

```
lines = sc.textFile("data.txt")
pairs = lines.map(lambda s:(s,1))
counts = pairs.reduceByKey(lambda a,b:a + b)
```

7.3.2 SQL引擎

Spark SQL可以使用JDBC/ODBC或命令行界面充当分布式查询引擎。在这种模式下，终端用户或应用程序可以直接与Spark SQL交互以运行SQL查询，而不需要编写任何代码。

1. 运行Thrift JDBC/ODBC服务器

这里实现的Thrift JDBC/ODBC服务器对应于Hive 1.2.1中的Hive Server 2。你可以使用Spark或Hive 1.2.1附带的脚本测试JDBC服务器。要启动JDBC/ODBC服务器，请在Spark目录中运行以下命令：

```
./sbin/start-thriftserver.sh
```

此脚本接受所有的bin/spark-submit命令行选项，--hiveconf选项用来指定Hive属性。可以运行./sbin/start-thriftserver.sh--help以获取所有可用选项的完整列表。默认情况下，服务器侦听localhost:10000。可以通过任一环境变量覆盖这个行为，实现命令如下所示：

```
export HIVE_SERVER2_THRIFT_PORT=<listening-port>
export HIVE_SERVER2_THRIFT_BIND_HOST=<listening-host>
./sbin/start-thriftserver.sh\
  --master <master-uri>\
```

2. 系统属性

系统属性命令如下所示：

```
./sbin/start-thriftserver.sh\
  --hiveconf hive.server2.thrift.port=<listening-port>\
  --hiveconf hive.server2.thrift.bind.host=<listening-host>\
  --master <master-uri>
```

3. 运行Spark SQL CLI

Spark SQL CLI是一种便捷的工具，可以在本地模式下运行Hive MetaStore服务并执行从命令行输入的查询。Spark SQL CLI无法与Thrift JDBC服务器通信。要启动Spark SQL CLI，请在Spark目录中运行以下命令：

```
./bin/spark-sql
```

可以将hive-site.xml、core-site.xml和hdfs-site.xml文件放在conf/下，可以运行./bin/spark-sql--help获取所有可用选项的完整列表。

7.4 PySpark案例

本节介绍的PySpark开发案例为官方社区开发的项目，版权为Apache基金会开源项目所有，无任何商业目的，其目的是让读者更快地理解PySpark。在此项目案例中，两个重要的概念是RDD和DataFrame。RDD和DataFrame在集群节点中为不可变的数据集合，PySpark还不支持DataSet，两者的区别为DataFrame是以命名列的方式组织数据的，类似于Pandas，而RDD的每一行都是一个String。

微课：v7-3 项目案例分析介绍

7.4.1 聚类分析

商业上，聚类可以帮助市场分析人员从消费者数据库中区分出不同的消费者群体，并且概括出每一类消费者的消费模式或者习惯。聚类作为数据挖掘中的一个模块，可以作为一个单独的工具来发现数据库中分布的一些深层的信息，并且概括出每一类的特点，或者把注意力放在某一个特定的类上；聚类分析也可以作为数据挖掘算法中的预处理步骤。kmeans_example.py的实现代码如下：

```
# Licensed to the Apache Software Foundation (ASF) under one or more
# contributor license agreements.See the NOTICE file distributed with
# this work for additional information regarding copyright ownership.
# The ASF licenses this file to You under the Apache License, Version 2.0
# (the "License"); you may not use this file except in compliance with
# the License.You may obtain a copy of the License at
## http://www.apache.org/licenses/LICENSE-2.0
## Unless required by applicable law or agreed to in writing, software
# distributed under the License is distributed on an "AS IS" BASIS,
# WITHOUT WARRANTIES OR CONDITIONS OF ANY KIND, either express or implied.
# See the License for the specific language governing permissions and
# limitations under the License.
#
"""
The K-means algorithm written from scratch against PySpark.In practice,
one may prefer to use the K-means algorithm in ML,as shown in
examples/src/main/python/ml/kmeans_example.py.
This example requires NumPy (http://www.numpy.org/).
"""
import sys
import numpy as np
from pyspark.sql import SparkSession
def parseVector(line):
    return np.array([float(x) for x in line.split('')])
def closestPoint(p,centers):
    bestIndex = 0
    closest = float("+inf")
    for i in range(len(centers)):
```

```
        tempDist = np.sum((p - centers[i]) ** 2)
        if tempDist < closest:
            closest = tempDist
            bestIndex = i
    return bestIndex
if _ _name_ _ == "_ _main_ _":
    if len(sys.argv) != 4:
        print("Usage: kmeans <file> <k> <convergeDist>",file=sys.stderr)
        sys.exit(-1)
    print("""WARN:This is a naive implementation of KMeans Clustering
        and is given
        as an example!Please refer to examples/src/main/python/ml/
            kmeans_example.py for an
        example on how to use ML’s KMeans implementation.""",
            file=sys.stderr)
    spark = SparkSession\
        .builder\
        .appName("PythonKMeans")\
        .getOrCreate()
    lines = spark.read.text(sys.argv[1]).rdd.map(lambda r:r[0])
    data = lines.map(parseVector).cache()
    K = int(sys.argv[2])
    convergeDist = float(sys.argv[3])
    kPoints = data.takeSample(False,K,1)
    tempDist = 1.0
    while tempDist > convergeDist:
        closest = data.map(
            lambda p:(closestPoint(p,kPoints),(p,1)))
        pointStats = closest.reduceByKey(
            lambda p1_c1,p2_c2:(p1_c1[0] + p2_c2[0],p1_c1[1] + p2_c2[1]))
        newPoints = pointStats.map(
            lambda st:(st[0],st[1][0] / st[1][1])).collect()
        tempDist = sum(np.sum((kPoints[iK] - p) ** 2) for (iK,p)
            in newPoints)
        for (iK,p) in newPoints:
            kPoints[iK] = p
    print("Final centers:" + str(kPoints))
    spark.stop()
```

7.4.2 数据处理

使用Apache Arrow助力PySpark数据处理。在大数据时代之前，大部分存储引擎使用的是按行存储的形式，很多早期的系统，如交易系统、ERP系统等，每次处理的是增、删、改、查某一个实体的所有信息，按行存储的话，能够快速地定位到单个实体并进行处理。随着大数据时代的到来，尤其是数据分析的不断发展，任务不需要一次读取实体的所有属性，只关心特定的某些属性，并对这些属性进行复杂的操作等。这种情况下，行存储需要读取额外的数据，这样就形成了瓶颈。而列存储将会减少额外数据的读取，还可以对相同属性的数据进行压缩，大大加快了处理速度。实现代码如下所示：

```
# Licensed to the Apache Software Foundation (ASF) under one or more
# contributor license agreements.See the NOTICE file distributed with
# this work for additional information regarding copyright ownership.
# The ASF licenses this file to You under the Apache License,Version 2.0
# (the "License");you may not use this file except in compliance with
# the License.You may obtain a copy of the License at
# http://www.apache.org/licenses/LICENSE-2.0
# Unless required by applicable law or agreed to in writing,software
# distributed under the License is distributed on an "AS IS" BASIS,
# WITHOUT WARRANTIES OR CONDITIONS OF ANY KIND,either express or
implied.
# See the License for the specific language governing permissions and
# limitations under the License.
#
"""
A simple example demonstrating Arrow in Spark.
Run with:
    ./bin/spark-submit examples/src/main/python/sql/arrow.py
"""
# NOTE that this file is imported in user guide in PySpark documentation.
# The codes are referred via line numbers.See also `literalinclude`
directive in Sphinx.
import pandas as pd
from typing import Iterable
from pyspark.sql import SparkSession
from pyspark.sql.pandas.utils import require_minimum_pandas_version,
require_minimum_pyarrow_version
require_minimum_pandas_version()
require_minimum_pyarrow_version()
def dataframe_with_arrow_example(spark:SparkSession) -> None:
    import numpy as np
    import pandas as pd
    # Enable Arrow-based columnar data transfers
    spark.conf.set("spark.sql.execution.arrow.pyspark.enabled","true")
    # Generate a Pandas DataFrame
    pdf = pd.DataFrame(np.random.rand(100,3))
    # Create a Spark DataFrame from a Pandas DataFrame using Arrow
    df = spark.createDataFrame(pdf)
    # Convert the Spark DataFrame back to a Pandas DataFrame using Arrow
    result_pdf = df.select("*").toPandas()
    print("Pandas DataFrame result statistics:\n%s\n" % str(
        result_pdf.describe()))
def ser_to_frame_pandas_udf_example(spark:SparkSession) -> None:
    import pandas as pd
    from pyspark.sql.functions import pandas_udf
    @pandas_udf("col1 string,col2 long") # type:ignore[call-overload]
    def func(s1:pd.Series,s2:pd.Series,s3:pd.DataFrame) -> pd.DataFrame:
        s3['col2'] = s1 + s2.str.len()
        return s3
    # Create a Spark DataFrame that has three columns including a
      struct column.
    df = spark.createDataFrame(
```

```
        [[1,"a string",("a nested string",)]],
        "long_col long,string_col string,struct_col struct<col1:string>")
    df.printSchema()
    # root
    # |-- long_column:long (nullable = true)
    # |-- string_column:string (nullable = true)
    # |-- struct_column:struct (nullable = true)
    # |    |-- col1:string (nullable = true)

    df.select(func("long_col","string_col","struct_col")).printSchema()
    # |-- func(long_col,string_col,struct_col):struct (nullable = true)
    # |    |-- col1:string (nullable = true)
    # |    |-- col2:long (nullable = true)
def ser_to_ser_pandas_udf_example(spark:SparkSession) -> None:
    import pandas as pd
    from pyspark.sql.functions import col,pandas_udf
    from pyspark.sql.types import LongType
    # Declare the function and create the UDF
    def multiply_func(a:pd.Series,b:pd.Series) -> pd.Series:
        return a * b
    multiply = pandas_udf(multiply_func,returnType=LongType()) # type:
        ignore[call-overload]
    # The function for a pandas_udf should be able to execute with
        local Pandas data
    x = pd.Series([1,2,3])
    print(multiply_func(x,x))
    # 0    1
    # 1    4
    # 2    9
    # dtype:int64
    # Create a Spark DataFrame,'spark' is an existing SparkSession
    df = spark.createDataFrame(pd.DataFrame(x,columns=["x"]))
    # Execute function as a Spark vectorized UDF
    df.select(multiply(col("x"),col("x"))).show()
    # +-------------------+
    # |multiply_func(x, x)|
    # +-------------------+
    # |                  1|
    # |                  4|
    # |                  9|
    # +-------------------+
def iter_ser_to_iter_ser_pandas_udf_example(spark:SparkSession) -> None:
    from typing import Iterator
    import pandas as pd
    from pyspark.sql.functions import pandas_udf
    pdf = pd.DataFrame([1,2,3],columns=["x"])
    df = spark.createDataFrame(pdf)
    # Declare the function and create the UDF
    @pandas_udf("long") # type:ignore[call-overload]
    def plus_one(iterator:Iterator[pd.Series]) -> Iterator[pd.Series]:
        for x in iterator:
            yield x + 1
```

```
    df.select(plus_one("x")).show()
    # +-----------+
    # |plus_one(x)|
    # +-----------+
    # |          2|
    # |          3|
    # |          4|
    # +-----------+
def iter_sers_to_iter_ser_pandas_udf_example(spark:SparkSession) -> None:
    from typing import Iterator,Tuple
    import pandas as pd
    from pyspark.sql.functions import pandas_udf
    pdf = pd.DataFrame([1,2,3],columns=["x"])
    df = spark.createDataFrame(pdf)
    #Declare the function and create the UDF
    @pandas_udf("long") # type:ignore[call-overload]
    def multiply_two_cols(
            iterator:Iterator[Tuple[pd.Series,pd.Series]]) ->
                Iterator[pd.Series]:
        for a,b in iterator:
            yield a * b
    df.select(multiply_two_cols("x","x")).show()
    # +-----------------------+
    # |multiply_two_cols(x,x)|
    # +-----------------------+
    # |                      1|
    # |                      4|
    # |                      9|
    # +-----------------------+
def ser_to_scalar_pandas_udf_example(spark:SparkSession) -> None:
    import pandas as pd
    from pyspark.sql.functions import pandas_udf
    from pyspark.sql import Window
    df = spark.createDataFrame(
        [(1,1.0),(1,2.0),(2,3.0),(2,5.0),(2,10.0)],
        ("id","v"))
    # Declare the function and create the UDF
    @pandas_udf("double") # type:ignore[call-overload]
    def mean_udf(v:pd.Series) -> float:
        return v.mean()
    df.select(mean_udf(df['v'])).show()
    # +-----------+
    # |mean_udf(v)|
    # +-----------+
    # |        4.2|
    # +-----------+
    df.groupby("id").agg(mean_udf(df['v'])).show()
    # +---+-----------+
    # | id|mean_udf(v)|
    # +---+-----------+
    # |  1|        1.5|
    # |  2|        6.0|
```

```
    # +---+-----------+
    w = Window\
        .partitionBy('id')\
        .rowsBetween(Window.unboundedPreceding,Window.unboundedFollowing)
    df.withColumn('mean_v',mean_udf(df['v']).over(w)).show()
    # +---+----+------+
    # | id|   v|mean_v|
    # +---+----+------+
    # |  1| 1.0|   1.5|
    # |  1| 2.0|   1.5|
    # |  2| 3.0|   6.0|
    # |  2| 5.0|   6.0|
    # |  2|10.0|   6.0|
    # +---+----+------+
def grouped_apply_in_pandas_example(spark:SparkSession) -> None:
    df = spark.createDataFrame(
        [(1,1.0),(1,2.0),(2,3.0),(2,5.0),(2,10.0)],
        ("id","v"))
    def subtract_mean(pdf:pd.DataFrame) -> pd.DataFrame:
        # pdf is a pandas.DataFrame
        v = pdf.v
        return pdf.assign(v=v - v.mean())
    df.groupby("id").applyInPandas(subtract_mean,schema=
        "id long,v double").show()
    # +---+----+
    # | id|   v|
    # +---+----+
    # |  1|-0.5|
    # |  1| 0.5|
    # |  2|-3.0|
    # |  2|-1.0|
    # |  2| 4.0|
    # +---+----+
def map_in_pandas_example(spark:SparkSession) -> None:
    df = spark.createDataFrame([(1,21),(2,30)],("id","age"))
    def filter_func(iterator:Iterable[pd.DataFrame]) -> Iterable[pd.
DataFrame]:
        for pdf in iterator:
            yield pdf[pdf.id == 1]
    df.mapInPandas(filter_func,schema=df.schema).show()
    # +---+---+
    # | id|age|
    # +---+---+
    # |  1| 21|
    # +---+---+
def cogrouped_apply_in_pandas_example(spark:SparkSession) -> None:
    import pandas as pd
    df1 = spark.createDataFrame(
        [(20000101,1,1.0),(20000101,2,2.0),(20000102,1,3.0),(20000102,2,4.0)],
        ("time","id","v1"))
    df2 = spark.createDataFrame(
        [(20000101,1,"x"),(20000101,2,"y")],
```

```
        ("time","id","v2"))
    def asof_join(left:pd.DataFrame,right:pd.DataFrame) -> pd.DataFrame:
        return pd.merge_asof(left,right,on="time",by="id")
    df1.groupby("id").cogroup(df2.groupby("id")).applyInPandas(
        asof_join,schema="time int,id int,v1 double,v2 string").show()
    # +--------+---+---+---+
    # |    time| id| v1| v2|
    # +--------+---+---+---+
    # |20000101|  1|1.0|  x|
    # |20000102|  1|3.0|  x|
    # |20000101|  2|2.0|  y|
    # |20000102|  2|4.0|  y|
    # +--------+---+---+---+
if __name__ == "__main__":
    spark = SparkSession \
        .builder \
        .appName("Python Arrow-in-Spark example") \
        .getOrCreate()
    print("Running Pandas to/from conversion example")
    dataframe_with_arrow_example(spark)
    print("Running pandas_udf example:Series to Frame")
    ser_to_frame_pandas_udf_example(spark)
    print("Running pandas_udf example:Seriesto Series")
    ser_to_ser_pandas_udf_example(spark)
    print("Running pandas_udf example:Iterator of Series to
        Iterator of Series")
    iter_ser_to_iter_ser_pandas_udf_example(spark)
    print("Running pandas_udf example:Iterator of Multiple Series to
        Iterator of Series")
    iter_sers_to_iter_ser_pandas_udf_example(spark)
    print("Running pandas_udf example:Series to Scalar")
    ser_to_scalar_pandas_udf_example(spark)
    print("Running pandas function example:Grouped Map")
    grouped_apply_in_pandas_example(spark)
    print("Running pandas function example:Map")
    map_in_pandas_example(spark)
    print("Running pandas function example:Co-grouped Map")
    cogrouped_apply_in_pandas_example(spark)
    spark.stop()
```

7.4.3 PageRank算法

PageRank算法的核心思想是数学支撑，也称网页排名、谷歌左侧排名，是一种由搜索引擎根据网页之间相互的超链接计算的技术，而作为网页排名的要素之一，以Google公司创办人拉里·佩奇（Larry Page）之姓来命名。Google公司用其来体现网页的相关性和重要性，在搜索引擎优化操作中，经常用来评估网页优化的成效因素，代码如下所示：

```
# the License.You may obtain a copy of the License at
#
```

```
#    http://www.apache.org/licenses/LICENSE-2.0
#
# Unless required by applicable law or agreed to in writing,software
# distributed under the License is distributed on an "AS IS" BASIS,
# WITHOUT WARRANTIES OR CONDITIONS OF ANY KIND,either express or implied.
# See the License for the specific language governing permissions and
# limitations under the License.
#
"""
This is an example implementation of PageRank.For more conventional use,
Please refer to PageRank implementation provided by graphx
Example Usage:
bin/spark-submit examples/src/main/python/pagerank.py data/mllib/
pagerank_data.txt 10
"""
import re
import sys
from operator import add
from typing import Iterable,Tuple
from pyspark.resultiterable import ResultIterable
from pyspark.sql import SparkSession
def computeContribs(urls:ResultIterable[str],rank:float) -> Iterable
[Tuple[str,float]]:
    """Calculates URL contributions to the rank of other URLs."""
    num_urls = len(urls)
    for url in urls:
        yield (url,rank/num_urls)
def parseNeighbors(urls:str) -> Tuple[str,str]:
    """Parses a urls pair string into urls pair."""
    parts = re.split(r '\s+',urls)
    return parts[0],parts[1]
if _ _name_ _ == "_ _main_ _":
    if len(sys.argv) != 3:
        print("Usage:pagerank <file> <iterations>",file=sys.stderr)
        sys.exit(-1)
    print("WARN:This is a naive implementation of PageRank and is given
as an example!\n" +
          "Please refer to PageRank implementation provided by graphx",
          file=sys.stderr)
    # Initialize the spark context.
    spark = SparkSession\
        .builder\
        .appName("PythonPageRank")\
        .getOrCreate()
    # Loads in input file.It should be in format of:
    #     URL         neighbor URL
    #     URL         neighbor URL
    #     URL         neighbor URL
    #     ...
    lines = spark.read.text(sys.argv[1]).rdd.map(lambda r:r[0])
    # Loads all URLs from input file and initialize their neighbors.
    links = lines.map(lambda urls: parseNeighbors(urls)).distinct().
```

```
groupByKey().cache()
    # Loads all URLs with other URL(s) link to from input file and initialize
      ranks of them to one.
    ranks = links.map(lambda url_neighbors:(url_neighbors[0],1.0))
    # Calculates and updates URL ranks continuously using PageRank
      algorithm.
    for iteration in range(int(sys.argv[2])):
        # Calculates URL contributions to the rank of other URLs.
        contribs = links.join(ranks).flatMap(lambda url_urls_rank:
            computeContribs(
            url_urls_rank[1][0],url_urls_rank[1][1] # type:ignore[arg-type]
        ))
        # Re-calculates URL ranks based on neighbor contributions.
        ranks = contribs.reduceByKey(add).mapValues(
            lambda rank:rank * 0.85 + 0.15)
    # Collects all URL ranks and dump them to console.
    for (link,rank) in ranks.collect():
        print("%s has rank:%s." % (link,rank))
    spark.stop()
```

7.5 本章小结

本章首先介绍了Apache Spark社区开发的一款工具PySpark的主要功能，Spark是实现API的一个计算框架。其次介绍了利用PySpark中的Py4j库，可以通过Python语言操作RDD。最后详细介绍了PySpark开发环境的部署与安装。

7.6 课后习题

1.选择题

（1）Apache Spark是用（　）语言实现的计算框架。

A.Java　　B.Python　　C.Scala　　D.C

（2）以下不为PySpark的常用API的是（　）。

A.Spark SQL　　B.DataFrame　　C.Streaming　　D.MSSQL

（3）Spark 3.2.1适用于Python版本的是（　）。

A.Python 3.6+　　B.Python 3.4　　C.Python 2.7+　　D.Python 2.8

（4）启动PySpark命令的是（　）。

A.start　　B.spark　　C.pstart　　D.pyspark

（5）开发PySpark程序使用的默认语言是（　）。

A.Java　　B.Python　　C.Scala　　D.C++

2.简答题

（1）简述PySpark的作用及其特点。

（2）简述PySpark常用的编程API及其功能。

（3）简述ApacheSpark与Hadoop的关系。

第8章 Flink开发应用

学习目标

（1）通过本章的学习，可让初学者了解Apache Flink技术。

（2）由浅入深地认识Apache Flink的概念。

（3）了解Apache Flink原理、安装部署、常用API、流数据项目开发的知识。

（4）为初学者打下基础。更进一步了解目前主流处理架构，以及未来流处理技术的发展，带领初学者走进流处理技术。

思政目标

（1）发展乡村产业是振兴乡村经济的根本所在，要在做大做强特色农业的基础上，大力依托农业农村资源的二三产业，智慧农业带动农业产业链实现全新变革。

（2）培养学生树立服务意识，学习好Flink流处理技术，为乡村智慧农业提供服务。

8.1 Flink概述

8.1.1 Flink简介

Apache Flink是由Apache软件基金会开发的开源流处理框架，其核心是使用Java语言和Scala语言编写的分布式流数据、流引擎。Flink以数据并行和流水线的方式执行任意流数据程序，Flink的流水线运行时，系统可以执行批处理和流处理程序。此外，Flink的运行时本身也支持迭代算法。

1. 数据流的运行流程

Flink程序在执行后会被映射到流数据、流中，每个Flink数据流以一个或多个源数据输入，例如从消息队列或文件系统开始，并以一个或多个接收器数据输出，如到消息队列文件系统或数据库等结束。Flink可以对流进行任意数量的变换，这些流可以被编排为有向无环数据流图，允许应用程序分支和合并数据流。

2. Flink的数据源和接收器

Flink提供现成的源和接收连接器，包括Apache Kafka Amazon Kinesis HDFS和Apache Cassandra等。Flink程序可以作为集群内的分布式系统运行，也可以独立的模式或在YARN Mesos基于Docker的环境和其他资源管理框架下进行部署。

3. Flink的状态、检查点和容错

Flink检查点是应用程序状态和源流中位置的自动异步快照机制。在发生故障的情况下，启用检查点的Flink程序将在恢复时从上一个完成的检查点恢复处理，确保Flink在应用程序中保持一次性状态语义。检查点机制暴露应用程序代码的接口，以便将外部系统包括在检查点机制中，例如打开和提交数据库系统的事务。Flink保存点的机制是一种手动触发的检查点。用户可以生成保存点，停止正在运行的Flink程序，然后从流中的相同应用程序状态和位置恢复程序。保存点可以在不丢失应用程序状态的情况下对Flink程序或Flink群集进行更新。

4. Flink的数据流API

Flink的数据流API支持有界或无界数据流上的转换，如过滤器聚合和窗口函数包含20多种不同类型的转换，可以在Java和Scala中使用。例如，一个简单的Scala有状态流处理程序，是从连续输入流发出字数，并在窗口中5秒对数据进行分组的应用。Apache Beam“提供了一种高级统一编程模型，允许开发人员实现可在在任何执行引擎上运行批处理和流数据处理作业”。Apache Flink-on-Beam运行器是功能最丰富的由Beam社区维护的能力矩阵。

5. 数据集API

Flink的数据集API支持对有界数据集进行转换，如过滤、映射、连接和分组，包含20多种不同类型的转换。该API可用于Java Scala和实验性的Python API。Flink的数据集API在概念上与数据流API类似。Flink的表API是一种类似SQL的表达式语言，用于关系流和批处理，可以嵌入Flink的Java、Scala数据集和数据流API中。表API和SQL接口在关系表抽象上运行，可以从外部数据源或现有数据流和数据集创建表。表API支持关系运算符，如表上的选择、聚合和连接等。也可以使用常规SQL查询表。表API提供了与SQL相同的功能，可以在同一程序中混合使用。将表转换回数据集或数据流时，由关系运算符和SQL查询定义的逻辑计划将使用Apache Calcite进行优化，并转换为数据集或数据流程序。

8.1.2 Flink与电商

Apache Flink作为开源流处理框架，在我国电子商务飞速发展的今天，Flink流处理技术时刻影响着我们的电商领域。伴随着国内电商行业的飞速发展，原有的“大范围”“广撒网”的运营方式已经不能满足业务需求。通过实时计算来快速挖掘用户的特征，分析用户的需求喜好，帮助电商企业完成数字化、精细化、个性化运营的方向转型，这就是Flink独特的核心技术。谈到电商，不得不谈到阿里巴巴团队，由阿里巴巴开发的阿里云实时计算是一个基于Apache Flink构建的企业级、高性能实时大数据处理系统，广泛适用于流式数据处理、离线数据处理等场景。阿里云实时计算完全兼容开源Flink API，能提供丰富的企业级增值功能。下面介绍阿里云实时计算核心功能。

1. 实时数据

协助企业建立数字化运营体系，实时统计各项关键业务指标，直观呈现企业生产经营情况，为经营决策提供实时数据支持。

2. 实时ETL

内置集成数十种连接器，覆盖数据库、消息队列、OLAP引擎等系统。使用全托管服务对数据进行实时流转集成，帮助企业构建数据平台。

3. 实时反作弊

从海量数据中实时识别刷单作弊、恶意爬虫等业务风险，避免企业出现巨大的经济损失，还可借助CEP（复杂事件处理）直接在流式处理作业中进行异常情况的检测。

4. 实时监测

实时计算Flink版高效的状态管理、丰富的窗口支持等特点可以帮助企业简化规则告警配置流程、提高告警效果，降低监测平台的维护成本。

5. 实时推荐

实时分析用户行为、结合AI技术建立更加精准的用户画像，及时推荐给用户更适合的新闻、视频和商品。

6. 实时IoT数据分析

实时捕捉、分析IoT设备产生的巨量数据，帮助用户实时分析和诊断设备的运行状况，实时检测运行时故障，实时预测制品良率等。Flink实时计算架构如图8-1所示。

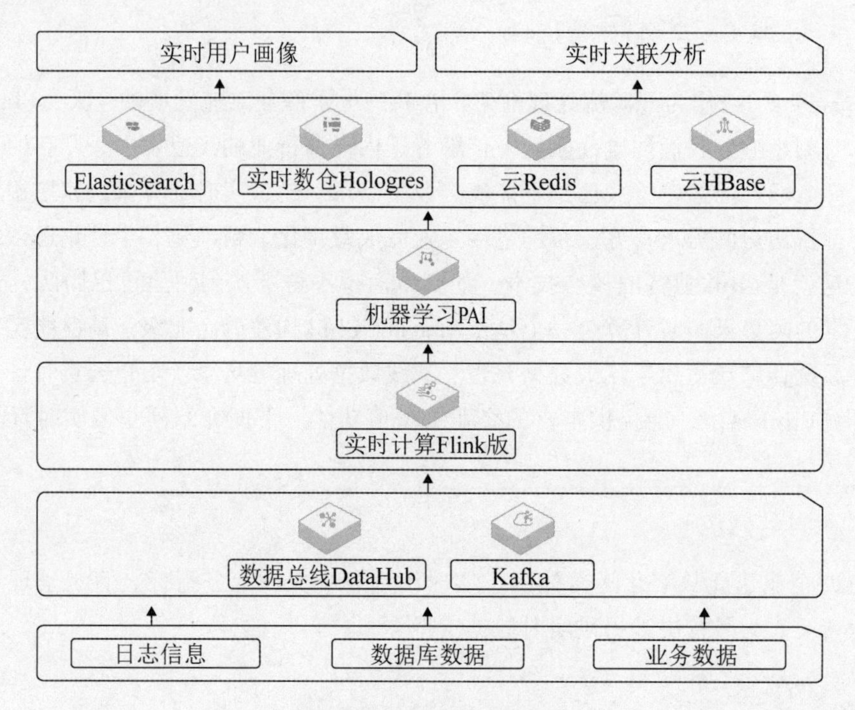

图8-1 Flink实时计算架构

通过以上对Flink实时计算的介绍，让我们重新认识了电商平台实时计算功能的强大。表面是电商平台改变了我们的生活习惯，为每一位人民群众提供了便利，其实是新的核心实时计算技术在改变我们的生活。核心技术的背后，实质上是强大的团队在共同参与Flink实时计算技术的开发。虽然Apache Flink为开源社区技术项目，但在我国的应用领域比较广泛，因此成就了一大批爱好者追逐此技术。

8.2 Flink部署

Flink可以运行在Linux、MacOSX和Windows操作系统上。在大型应用生产系统中，以Linux操作系统较多。作为一个开发人员，如果进行本地开发测试，那么采用本地模式的安装比较方便。Flink对系统要求不是特别严格，唯一要求是Java 1.7.x或更高版本，本地运行会启动Single JVM，主要用于测试和调试代码。

8.2.1 Flink架构简介

Flink集群总是包含一个JobManager以及一个或多个TaskManager。JobManager负责处理、提交、监控以及资源管理。TaskManager运行worker进程，负责实际任务的执行，而这些任务共同组成一个Flink Job。这里，我们会先运行一个TaskManager，然后扩容到多个TaskManager。另外，我们会专门使用一个客户端容器来提交Flink Job，后续还会使用该容器执行一些操作。需要注意的是，Flink集群的运行并不需要依赖客户端容器，这里引入只是为了使用方便。Kafka集群由一个Zookeeper服务端和一个Kafka Broker组成。Flink组成图如图8-2所示。

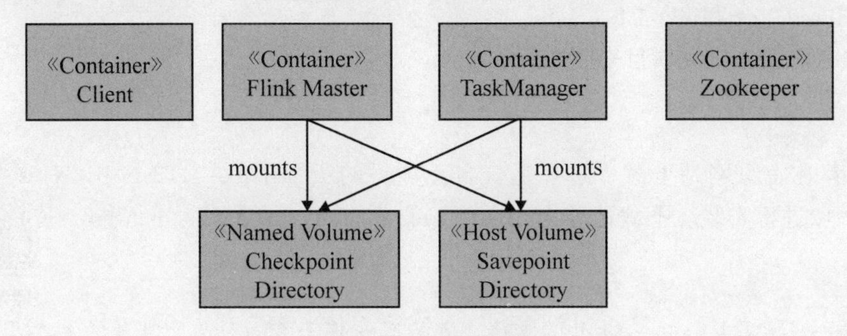

图8-2 Flink组成图

8.2.2 输入流程

数据输入时会往JobManager提交一个名为Flink事件计数的Job，还会创建两个Kafka Topic：input和output。该Job负责从input Topic消费点击事件（Click Event），每个点击事件都包含一个timestamp和一个page属性。这些事件将按照page属性进行分组，然后按照每15 s窗口进行统计，最终结果输出到output Topic中。共有6种不同的page属性。针对特定的page属性，我们会按照每15 s产生1000个点击事件的速率生成数据。因此，Flink Job应该能在每个窗口中输出1000个page属性的点击数据。输入/输出过程图如图8-3所示。

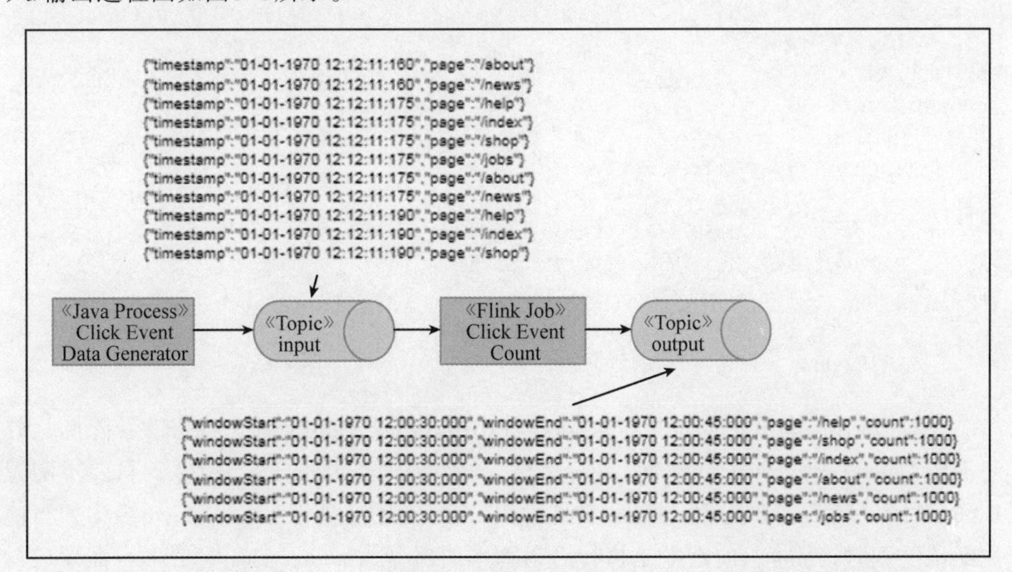

图8-3 输入/输出过程图

8.2.3 环境搭建

Flink环境的搭建比较简单，只需要几步就可以完成。实训环境采用虚拟主机形式进行介绍，读者可根据自己的环境如虚拟机、公有云服务器和私有云服务器来安装。

Flink软件支持linux-CentOS7以上的版本、hadoop-2.7.5以上的版本、Scala-2.11.6和jdk-1.8。

Flink可根据开发项目来制定版本。

1. Flink软件下载

推荐在官方网站下载，选择适合自己项目的版本flink-1.13.6-bin-scala_2.12进行安装。下载地址为https://flink.apache.org/downloads.html。

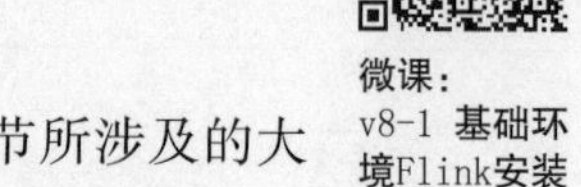

微课：
v8-1 基础环境Flink安装

2. 安装部署

Flink安装需要hadoop和jdk的支持，安装操作基于前面章节所涉及的大数据技术，其他软件的安装详情请参照前面章节的详细介绍。代码如下：

```
#解压软件
[root@master ~]# tar -zxvf flink-1.13.6-bin-scala_2.12.tgz
flink-1.13.6/conf/log4j.properties
flink-1.13.6/conf/logback-console.xml
flink-1.13.6/conf/masters
flink-1.13.6/conf/log4j-console.properties
flink-1.13.6/conf/zoo.cfg
flink-1.13.6/conf/logback.xml
flink-1.13.6/conf/log4j-session.properties
flink-1.13.6/conf/workers
flink-1.13.6/conf/logback-session.xml
flink-1.13.6/conf/flink-conf.yaml
flink-1.13.6/conf/log4j-cli.properties
flink-1.13.6/NOTICE
#复制安装文件
[root@master ~]# cp -r flink-1.13.6 /usr/local/flink-1.13.6
#环境变量配置
[root@master ~]#vi /etc/profile
#flink
export FLINK_HOME=/usr/local/flink-1.13.6
export PATH=$FLINK_HOME/bin:$PATH
[root@master ~]# source /etc/profile
```

3. 配置管理

在管理Flink常用配置时，需对slaves和flink-conf.yaml文件进行修改，除此之外还需要对masters文件进行配置。在masters文件中，主要用于配置HA，只要不配置HA的环境，就不需要配置masters文件。Flink配置也是基于master/slave结构，当使用默认配置时，master的选择是执行启动脚本的机器为master端。但是slave需要进行手动配置，配置对应的主机名即可修改为伪分布式和分布式。主要区别实际上取决于slave节点的个数，以及分布式在多个节点上而已。接下来介绍flink-conf.yaml配置与其他配置的不同点，在配置过程中Flink和Spark的比较配置格式有一定区别。Spark配置文件分spark-env.sh和spark-default.conf文件，而Flink的配置都在flink-conf.yaml配置文件中。详细配置代码如下所示：

```
#JobManager runs.
jobmanager.rpc.address: cdh1
#The RPC port where the JobManager is reachable.
jobmanager.rpc.port: 6123
#The heap size for the JobManager JVM
jobmanager.heap.size: 1024m
#The heap size for the TaskManager JVM
taskmanager.heap.size: 1024m
#The number of task slots that each TaskManager offers. Each slot runs
one parallel pipeline.
taskmanager.numberOfTaskSlots: 1
#The parallelism used for programs that did not specify and other
parallelism.
parallelism.default: 1
#配置是否在Flink集群启动时给TaskManager分配内存，默认不进行预分配，这样在我们不适
用Flink集群时不会占用集群资源
taskmanager.memory.preallocate: false
#用于未指定程序的并行性和其他并行性，默认并行度
parallelism.default: 2
#用于指定JobManger的可视化端口，尽量配置一个不容易冲突的端口
jobmanager.web.port: 5566
#配置checkpoint目录
state.backend.fs.checkpointdir: hdfs://cdh1:9000/flink-checkpoints
#配置hadoop的配置文件
fs.hdfs.hadoopconf: /usr/local/hadoop/etc/hadoop/
#访问hdfs系统使用的
fs.hdfs.hdfssite: /usr/local/hadoop/etc/hadoop/hdfs-site.xml
```

8.2.4 Flink Web用户界面介绍

在Flink Web用户界面打开浏览器并访问http://localhost:8081，如果一切正常，将会在Web界面上看到一个TaskManager和一个处于“RUNNING”状态的名为Click Event Count的Job。Flink Web用户界面包含许多关于Flink集群以及运行在其上的Jobs的有用信息，比如JobGraph Metrics Checkpointing Statistics TaskManager Status。Web用户界面如图8-4所示。

1. Job管理

Job提交后，Flink会默认为其生成一个JobID，后续再执行该Job的所有操作，无论是通过CLI还是通过REST API都需要带上JobID。在Job失败的情况下，Flink对事件处理依然能够提供一次保障。本节中你会观察到并能够在某种程度上验证这种行为。如前文所述，事件以特定速率生成，刚好使得每个统计窗口都包含确切的1000条记录。因此，你可以实时查看output Topic的输出，确定失败恢复后，所有的窗口依然输出正确的统计数字，以此来验证Flink在TaskManager失败时能够成功恢复，而且不丢失数据、不产生重复数据。为此，通过控制台命令消费output Topic，保持消费直到Job从失败中恢复。

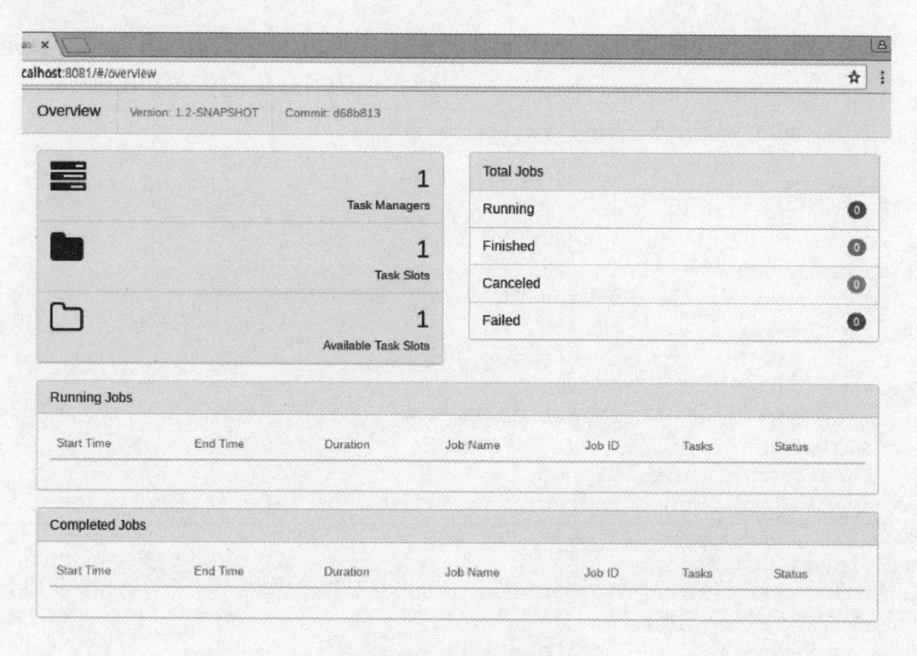

图8-4 Web用户界面

2. 模拟失败

为了模拟部分失败故障，可以通过kill掉一个TaskManager，这种失败行为在生产环境中相当于TaskManager进程挂掉、TaskManager机器宕机或者从框架或用户代码中抛出一个临时异常（例如，由于外部资源暂时不可用）而导致的失败。几秒钟后，JobManager就会感知到TaskManager已失联，接下来它会取消Job运行并且立即重新提交该Job以进行恢复。当Job重启后，所有的任务都会处于SCHEDULED状态，Job状态信息如图8-5所示。

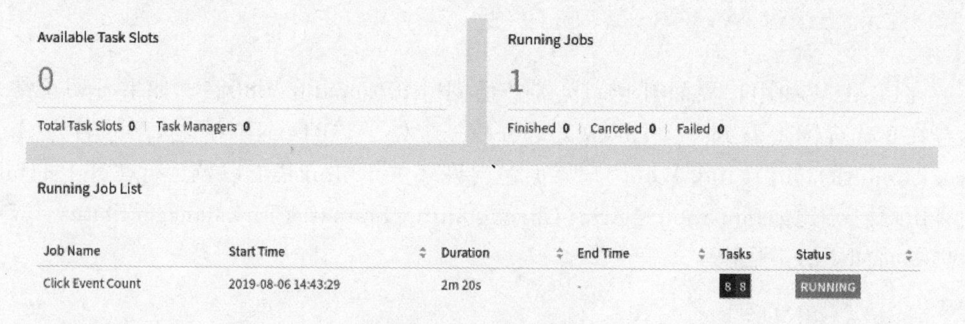

图8-5 Job状态信息

3. 重启Job管理

可以从这个Savepoint重新启动待升级的Job，为了简单操作，不对该Job作任何变更就直接重启。查询Job指标，可以通过JobManager提供的REST API来获取系统和用户指标，具体请求方式取决于我们想查询哪类指标。Job相关的指标分类可通过jobs/<job-id>/metrics获得，如果想要查询某类指标的具体值，则可以在请求地址后跟上get参数查询相应的信息。

8.3 Flink API

8.3.1 常用API介绍

微课：
v8-2 Flink
API介绍

Flink中的DataStream程序是对数据流进行转换的，例如过滤、更新状态、定义窗口、聚合的常规程序。数据流最初从各种来源,例如消息队列、套接字和流文件创建。结果通过接收器返回，例如，可以将数据写入文件或标准输出，例如命令行终端。Flink程序可以在各种上下文中运行，可以独立运行，也可以嵌入其他程序中。执行可以发生在本地JVM中，也可以发生在多台机器的集群上。为了创建自己的Flink DataStream程序，从剖析Flink程序开始，并逐渐添加认识流的转换。

数据流DataStream API得名于一个特殊的DataStream类，该类用于表示Flink程序中的数据集合，可以将它们视为包含重复项的不可变数据集合。这些数据可以是有限的，也可以是无限的，用于处理它们的API是相同的。在用法上，DataStream与常规的Java相似，Collection但在某些关键方面却大不相同。它们是不可变的，这意味着一旦它们被创建，就不能添加或删除元素。你也不能简单地检查内部元素，而只能使用DataStream API操作，也称转换处理它们。DataStream可以通过在Flink程序中添加源来创建首字母。然后，你可以从中派生出新的流，并使用API方法，例如map等将它们组合起来过滤。

Flink程序看起来像转换DataStreams。每个程序由相同的基本部分组成。有关更多详细信息请参阅相应部分。请注意，Java DataStream API的所有核心类都可以在org.apache.flink.streaming.api中找到。在应用程序开发过程中，StreamExecutionEnvironment是所有Flink程序的基础。可以使用以下静态方法获得一个StreamExecutionEnvironment，实现代码如下所示：

```
getExecutionEnvironment()
createLocalEnvironment()
createRemoteEnvironment(String host,int port,String...jarFiles)
```

通常，只需要使用getExecutionEnvironment()，因为将根据上下文执行正确的操作：如果在IDE中执行你的程序或作为常规Java程序，它将创建一个本地环境，该环境将在你的本地机器上执行你的程序。如果从你的程序中创建一个JAR文件，并通过命令行调用它，Flink集群管理器将执行getExecutionEnvironment()方法并返回一个执行环境来在集群上执行你的程序。为了指定数据源，执行环境可以使用几种方法从文件中读取：可以逐行读取它们，作为CSV文件，或使用其他提供的源。要将文本文件作为一系列行读取，即可以使用，实现代码如下所示：

```
final StreamExecutionEnvironment env = StreamExecutionEnvironment.
getExecutionEnvironment();
DataStream<String> text = env.readTextFile("file:///path/to/file");
```

通过类来生成DataStream，然后应用转换transformation来创建新的派生DataStream。可以调用DataStream上具有转换功能的方法来应用转换，如map的转换代码如下所示：

```
DataStream<String> input = ...;
DataStream<Integer> parsed = input.map(new MapFunction<String,Integer>() {
    @Override
    public Integer map(String value) {
        return Integer.parseInt(value);
    }
});
```

程序执行的最后一部分对于何时理解以及如何执行Flink操作至关重要。所有Flink程序都是惰性执行的，当程序执行main()方法时，数据加载和转换不会直接发生。相反，每个操作都会被创建并添加到数据流图中。当执行是由execute()对执行环境的调用显式触发时，这些操作才会真正执行。程序是在本地执行还是在集群上执行取决于执行环境的类型，Flink作为一个整体规划的单元执行。以下代码是一个完整的流窗口字数统计应用程序的示例。

```
import org.apache.flink.api.common.functions.FlatMapFunction;
import org.apache.flink.api.java.tuple.Tuple2;
import org.apache.flink.streaming.api.datastream.DataStream;
import org.apache.flink.streaming.api.environment.
StreamExecutionEnvironment;
import org.apache.flink.streaming.api.windowing.time.Time;
import org.apache.flink.util.Collector;
public class WindowWordCount {
    public static void main(String[] args) throws Exception {
        StreamExecutionEnvironment env =
            StreamExecutionEnvironment.getExecutionEnvironment();
        DataStream<Tuple2<String, Integer>> dataStream = env
                .socketTextStream("localhost",9999)
                .flatMap(new Splitter())
                .keyBy(value -> value.f0)
                .window(TumblingProcessingTimeWindows.of(Time.seconds(5)))
                .sum(1);
        dataStream.print();
        env.execute("Window WordCount");
    }
    public static class Splitter implements FlatMapFunction<String,
        Tuple2<String,Integer>> {
        @Override
        public void flatMap(String sentence,
            Collector<Tuple2<String,Integer>> out) throws Exception {
            for (String word: sentence.split(" ")) {
                out.collect(new Tuple2<String,Integer>(word,1));
            }
        }
    }
}
```

8.3.2 Watermark策略

为了使用事件时间语义，Flink应用程序需要知道事件时间戳对应的字段，意味着数据流中的每个元素都需要拥有可分配的事件时间戳。其通常通过使用TimestampAssigner API从元素中的某个字段去访问/提取时间戳。时间戳的分配与Watermark的生成齐头并进，这样就可以告诉Flink应用程序事件时间的进度。可以通过指定WatermarkGenerator来配置Watermark的生成方式。使用Flink API时，需要设置一个同时包含TimestampAssigner和WatermarkGenerator的WatermarkStrategy。WatermarkStrategy工具类中也提供了许多常用的Watermark策略，并且用户也可以在某些必要场景下构建自己的watermark策略，WatermarkStrategy接口程序代码如下所示：

```
public interface WatermarkStrategy<T>
    extends TimestampAssignerSupplier<T>,WatermarkGeneratorSupplier<T>{
    /**
     * 根据策略实例化一个可分配时间戳的{@link TimestampAssigner}。
     */
    @Override
    TimestampAssigner<T> createTimestampAssigner
        (TimestampAssignerSupplier.Context context);
    /**
     * 根据策略实例化一个watermark生成器
     */
    @Override
    WatermarkGenerator<T> createWatermarkGenerator
(WatermarkGeneratorSupplier.Context context);
}
```

使用Watermark策略，WatermarkStrategy可以在Flink应用程序中的两处使用，第一种是直接在数据源上使用，第二种是直接在非数据源的操作之后使用。相对来说，第一种方式更好，因为数据源可以利用Watermark生成逻辑中有关分片/分区的信息。使用这种方式，数据源通常可以更精准地跟踪Watermark，Watermark生成将更精确。直接在源上指定WatermarkStrategy，意味着你必须使用特定数据源接口。请参阅Watermark策略与Kafka连接器来了解如何使用Kafka Connector，以及有关每个分区的Watermark是如何生成以及工作的。实现的Java接口代码如下所示：

```
final StreamExecutionEnvironment env =
    StreamExecutionEnvironment.getExecutionEnvironment();
DataStream<MyEvent> stream = env.readFile(
    myFormat,myFilePath,FileProcessingMode.PROCESS_CONTINUOUSLY,100,
    FilePathFilter.createDefaultFilter(),typeInfo);
DataStream<MyEvent> withTimestampsAndWatermarks = stream
    .filter( event -> event.severity() == WARNING )
    .assignTimestampsAndWatermarks(<watermark strategy>);
withTimestampsAndWatermarks
    .keyBy((event) -> event.getGroup())
    .window(TumblingEventTimeWindows.of(Time.seconds(10)))
    .reduce((a,b) -> a.add(b))      .addSink(...);
```

自定义WatermarkGenerator，TimestampAssigner是一个可以从事件数据中提取时间戳字段的简单函数，我们无须详细查看其实现。但是，WatermarkGenerator的编写相对就要复杂一些，我们将在接下来的两节中介绍如何实现此接口。WatermarkGenerator接口的实现代码如下所示：

```
/**
 * {@code WatermarkGenerator} 可以基于事件或者周期性地生成watermark。
 *
 * <p><b>注意：</b> WatermarkGenerator将以前互相独立的{@code
AssignerWithPunctuatedWatermarks}
 * 和{@code AssignerWithPeriodicWatermarks}一同包含进来。
 */
@Public
public interface WatermarkGenerator<T> {
    /**
     * 接收到一个事件数据就调用一次，可以检查或者记录事件的时间戳，或者基于事件数据
       本身去生成watermark。
     */
    void onEvent(T event,long eventTimestamp,WatermarkOutput output);
    /**
     * 周期性调用，也许会生成新的watermark，也许不会。
     *
     * <p>调用此方法生成watermark的间隔时间由{@link
       ExecutionConfig#getAutoWatermarkInterval()}决定。
     */
    void onPeriodicEmit(WatermarkOutput output);
}
```

自定义周期性Watermark生成器，周期性生成器会观察流事件数据并定期生成Watermark，其生成可能取决于流数据，或者完全基于处理时间。生成Watermark的时间间隔（每n毫秒）可以通过ExecutionConfig.setAutoWatermarkInterval指定。每次都会调用生成器的onPeriodicEmit()方法，如果返回的Watermark非空且值大于前一个Watermark，则将发出新的Watermark。注意Flink已经附带了BoundedOutOfOrdernessWatermarks，实现了WatermarkGenerator，其工作原理与下面的BoundedOutOfOrdernessGenerator相似，实现代码如下所示：

```
public class BoundedOutOfOrdernessGenerator implements
WatermarkGenerator<MyEvent> {
    private final long maxOutOfOrderness = 3500; //3.5秒
    private long currentMaxTimestamp;
    @Override
    public void onEvent(MyEvent event,long eventTimestamp,
        WatermarkOutput output) {
        currentMaxTimestamp = Math.max(currentMaxTimestamp,
            eventTimestamp);
    }
    @Override
    public void onPeriodicEmit(WatermarkOutput output) {
        //发出的watermark = 当前最大时间戳 - 最大乱序时间
        output.emitWatermark(new Watermark(
```

```
            currentMaxTimestamp - maxOutOfOrderness - 1));
    }
}
/**
 * 该生成器生成的Watermark滞后于处理时间固定量。它假定元素会在有限延迟后到达Flink
 */
public class TimeLagWatermarkGenerator implements
WatermarkGenerator<MyEvent> {
    private final long maxTimeLag = 5000; //5秒
    @Override
    public void onEvent(MyEvent event,long eventTimestamp,
WatermarkOutput output) {
        //处理时间场景下不需要实现
    }
    @Override
    public void onPeriodicEmit(WatermarkOutput output) {
        output.emitWatermark(new Watermark(
            System.currentTimeMillis() - maxTimeLag));
    }
}
```

自定义标记Watermark生成器，标记Watermark生成器观察流事件数据并在获取到带有Watermark信息的特殊事件元素时发出Watermark。以下是实现标记生成器的方法，当事件带有某个指定标记时，该生成器就会发出Watermark，实现代码如下所示：

```
public class PunctuatedAssigner implements WatermarkGenerator<MyEvent> {
    @Override
    public void onEvent(MyEvent event,long eventTimestamp,
WatermarkOutput output) {
        if (event.hasWatermarkMarker()) {
            output.emitWatermark(new Watermark(
                event.getWatermarkTimestamp()));
        }
    }
    @Override
    public void onPeriodicEmit(WatermarkOutput output) {
        //onEvent已经实现
    }
}
```

8.3.3 Keyed DataStream

键控数据流，如果要使用键控状态，首先要指定一个键，该键DataStream应该用于对状态进行分区以及流中的记录。你可以在Java/Scala API或key_by或Python API中使用DataStream，这将产生KeyedStream，然后允许使用键控状态的操作。键选择器函数将单个记录作为输入并返回该记录的键。密钥可以是任何类型，并且必须来自确定性计算。Flink的数据模型不是基于键-值对的。因此，你不需要将数据集类型以物理方式打包到键和值中。键是“虚拟的”，它们被定义为实际数据上的函数，以指导分组操作符。以下示例显示了一个键选择器函数，该函数仅返回对象的字段，实现代码如下所示：

```
//some ordinary POJO
public class WC {
  public String word;
  public int count;
  public String getWord() {return word;}}DataStream<WC> words = //
[...]KeyedStream<WC> keyed = words
  .keyBy(WC::getWord);
```

KeyedState接口提供不同类型状态的访问接口，这些状态都作用在当前输入数据的key下。这些状态仅在KeyedStream上使用，在Java/ScalaAPI上可以通过stream.keyBy得到KeyedStream，在Python API上可以通过stream.key_by得到KeyedStream。不同类型的状态，然后介绍如何使用它们。ValueState用于保存一个可以更新和检索的值如上所述，每个值都对应到当前的输入数据的key中，因此，算子接收到的每个key都可能对应一个。这个值可以通过update进行更新，通过Tvalue进行检索。

ListState用于保存一个元素的列表。可以往这个列表中追加数据，并在当前列表上进行检索。可以通过add或者addAll添加元素，通过Iterable get获得整个列表，还可以通过update覆盖当前的列表。ReducingState用于保存一个单值，表示添加到状态的所有值的聚合。接口与ListState类似，但使用add增加元素，会使用提供的ReduceFunction进行聚合。AggregatingState保留一个单值，表示添加到状态的所有值的聚合。与ReducingState相反,聚合类型可能与添加到状态的元素的类型不同。接口与ListState类似，但使用add添加的元素会用指定的AggregateFunction进行聚合。MapState维护了一个映射列表。你可以添加键-值对到状态中，也可以获得反映当前所有映射的迭代器。使用put或者putAll添加映射，使用get检索特定key。使用entries、keys和values分别检索映射键和值的可迭代视图，还可以通过isEmpty来判断是否包含任何键-值对。所有类型的状态还有一个clear方法，用于清除当前key下的状态数据，也就是当前输入元素的key。请牢记，这些状态对象仅用于与状态交互。状态本身不一定存储在内存中，还可能存储在磁盘或其他位置。另外需要牢记的是，从状态中获取的值取决于输入元素所代表的key。在不同的key上调用同一个接口，可能会得到不同的值。必须创建一个StateDescriptor，才能得到对应的状态句柄。这保存了状态名称，正如我们稍后将看到的，你可以创建多个状态，并且它们必须有唯一的名称，以便可以引用它们，状态所持有值的类型，并且可能包含用户指定的函数，例如ReduceFunction。根据不同的状态类型，可以创建ValueStateDescriptor、ListStateDescriptor、AggregatingStateDescriptor、ReducingStateDescriptor或MapStateDescriptor。状态通过RuntimeContext进行访问，因此只能在rich functions中获取相关信息。RichFunction中的RuntimeContext提供的代码如下所示：

```
ValueState<T>
getState(ValueStateDescriptor<T> ReducingState<T> getReducingState
(ReducingStateDescriptor<T>ListState<T> getListState(ListStateDescriptor
<T>AggregatingState<IN,OUT>getAggregatingState(AggregatingStateDescriptor
<IN,ACC,OUT>)MapState<UK,UV>getMapState(MapStateDescriptor<UK,UV>)
```

以下是FlatMapFunction的组合实现代码：

```
public class CountWindowAverage extends RichFlatMapFunction<Tuple2
    <Long,Long>,Tuple2<Long,Long>> {
    /**
     * The ValueState handle. The first field is the count,the second
       field a running sum.
     */
    private transient ValueState<Tuple2<Long,Long>> sum;
    @Override
    public void flatMap(Tuple2<Long,Long> input,Collector
        <Tuple2<Long,Long>> out) throws Exception {
        // access the state value
        Tuple2<Long,Long> currentSum = sum.value();
        //update the count
        currentSum.f0 += 1;
        //add the second field of the input value
        currentSum.f1 += input.f1;
        // update the state
        sum.update(currentSum);
        // if the count reaches 2,emit the average and clear the state
        if (currentSum.f0 >= 2) {
            out.collect(new Tuple2<>(input.f0,currentSum.f1/currentSum.f0));
            sum.clear();
        }
    }
    @Override
    public void open(Configuration config) {
        ValueStateDescriptor<Tuple2<Long, Long>> descriptor =
                new ValueStateDescriptor<>(
                        "average", //the state name
                        TypeInformation.of(new TypeHint<Tuple2
                        <Long,Long>>(){}), //type information
                        Tuple2.of(0L,0L));
                        //default value of the state,if nothing was set
        sum = getRuntimeContext().getState(descriptor);
    }
}
//this can be used in a streaming program like this (assuming we have
  a StreamExecutionEnvironment env)
env.fromElements(Tuple2.of(1L,3L),Tuple2.of(1L,5L),Tuple2.of(1L,7L),
Tuple2.of(1L,4L),Tuple2.of(1L,2L))
        .keyBy(value -> value.f0)
        .flatMap(new CountWindowAverage())
        .print();
//the printed output will be (1,4) and (1,5)
```

通过实现CheckpointedFunction接口来使用operator state。进行checkpoint时会调用snapshotState()。在开发过程中，用户自定义函数初始化时会调用initializeState()，初始化包括第一次自定义函数初始化和从之前的checkpoint恢复。因此，initializeState()不仅是定义不同状态类型初始化的地方，也包括状态恢复的逻辑。operator state以list的形式存在。这些状态是一个可序列化对象的集合列表，彼此独立，方便在改变并发后进行状态的重新分派。换句话说，这些对象是重新分配non-keyed state的最细粒度。根据状态的不同，访问方式也不同，程序实现代码如下所示：

```
public class BufferingSink
        implements SinkFunction<Tuple2<String,Integer>>,
                   CheckpointedFunction {
    private final int threshold;
    private transient ListState<Tuple2<String,Integer>> checkpointedState;
    private List<Tuple2<String,Integer>> bufferedElements;
    public BufferingSink(int threshold) {
        this.threshold = threshold;
        this.bufferedElements = new ArrayList<>();
    }
    @Override
    public void invoke(Tuple2<String,Integer> value,Context contex)
        throws Exception {
        bufferedElements.add(value);
        if (bufferedElements.size() >= threshold) {
            for (Tuple2<String,Integer> element:bufferedElements) {
                //send it to the sink
            }
            bufferedElements.clear();
        }
    }
    @Override
    public void snapshotState(FunctionSnapshotContext context)
        throws Exception {
        checkpointedState.clear();
        for (Tuple2<String,Integer> element:bufferedElements) {
            checkpointedState.add(element);
        }
    }
    @Override
    public void initializeState(FunctionInitializationContext context)
        throws Exception {
        ListStateDescriptor<Tuple2<String,Integer>> descriptor =
            new ListStateDescriptor<>(
                "buffered-elements",
                TypeInformation.of(new TypeHint<Tuple2<String,
                    Integer>>() {}));
        checkpointedState = context.getOperatorStateStore().
            getListState(descriptor);
        if (context.isRestored()) {
            for (Tuple2<String,Integer> element:checkpointedState.get()) {
                bufferedElements.add(element);
            }
        }
    }
}
```

initializeState方法会接收一个FunctionInitializationContext参数，用来初始化non-keyed state的容器。这些容器是一个ListState，用于在checkpoint时保存non-keyed state对象，实现代码如下所示：

```
ListStateDescriptor<Tuple2<String,Integer>> descriptor =
    new ListStateDescriptor<>(
        "buffered-elements",
        TypeInformation.of(new TypeHint<Tuple2<String,Integer>>() {}));
checkpointedState = context.getOperatorStateStore().getListState
    (descriptor);
```

相比其他算子，带状态的Source Function的数据源需要注意的东西更多。为了保证更新状态以及输出的原子性支持exactly-once语义，用户需要在发送数据前获取数据源的全局锁。程序代码如下所示：

```
public static class CounterSource
        extends RichParallelSourceFunction<Long>
        implements CheckpointedFunction {
    /** current offset for exactly once semantics */
    private Long offset = 0L;
    /** flag for job cancellation */
    private volatile boolean isRunning = true;
     /** 存储state的变量. */
    private ListState<Long> state;
     @Override
    public void run(SourceContext<Long> ctx) {
        final Object lock = ctx.getCheckpointLock();
        while (isRunning) {
            //output and state update are atomic
            synchronized (lock) {
                ctx.collect(offset);
                offset += 1;
            }
        }
    }
    @Override
    public void cancel() {
        isRunning = false;
    }
    @Override
    public void initializeState(FunctionInitializationContext context)
        throws Exception {
        state = context.getOperatorStateStore().getListState(
            new ListStateDescriptor<>(
            "state",
            LongSerializer.INSTANCE));
        //从已保存的状态中将offset恢复到内存中，在进行任务恢复的时候，也会调用此初
          始化状态的方法
        for (Long l:state.get()) {
            offset = l;
        }
```

```
    }
    @Override
    public void snapshotState(FunctionSnapshotContext context)
        throws Exception {
        state.clear();
        state.add(offset);
    }
}
```

8.4 Flink案例

Flink批处理程序作为流处理程序被执行，DataSet内部被视为数据流。流处理程序同样适用于批处理程序。操作分为无状态的操作与有状态的操作，无状态的操作不包含任何内部状态。处理事件时并不需要任何其他事件的信息，也不需要保存它自己的信息。无状态的操作易于并行，因为事件可以以它们到达的顺序、相互独立而被处理。当出现错误时，无状态的操作可以被简单地重新执行，从它丢失数据的点开始继续执行即可。

8.4.1 项目案例简介

本章节介绍在某智慧农业电商企业，作为技术负责人主要负责公司大量的数据流管理，如经常策划搞运营活动提高农产品的销售量。通过一个活动下来，农业电商平台会产生大量的实时数据。但经过细致排查，发现原来多数平台的都在促销，效果却没达到。作为智慧农业电商平，开发项目初衷是服务三农，为乡村振兴需求出发去研发。在实施过程中可以做一个实时的异常检测系统，监控用户的高危行为，及时发施降低损失。用户的行为经由App上报或Web日志记录下来，发送到一个消息队列里去；然后流计算订阅消息队列，过滤出感兴趣的行为，比如购买、领券、浏览等；流计算把这个行为特征化；流计算通过UDF调用外部一个风险模型，判断这次行为是否有问题；流计算里通过CEP功能，跨多条记录分析用户行为，整体识别是否有风险；综合风险模型和CEP的结果产出预警信息。

微课：v8-3 智慧农业电商项目介绍

在批处理和流处理、电商用户行为分析、数据源解析、项目模块划、分批处理主要操作大容量静态数据集，并在计算过程完成后返回结果。处理的是用一个固定时间间隔分组的数据点集合。批处理模式中使用的数据集通常符合下列特征，批处理数据集代表数据的有限集合，数据通常始终存储在某种类型的持久存储位置中，批处理操作通常是处理极为海量数据集的唯一方法，用户行业数据流程图如图8-6所示。

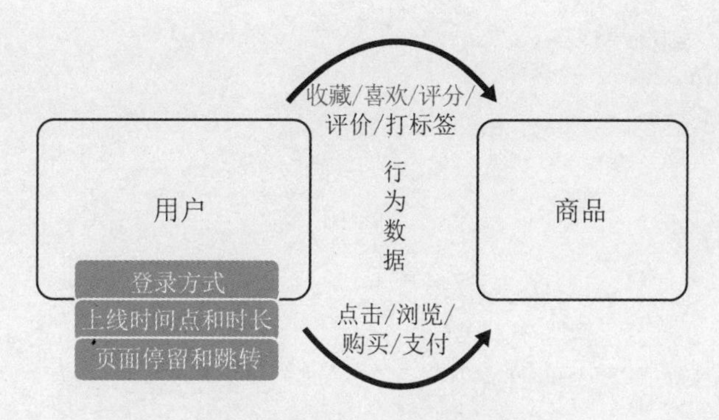

图8-6 用户行业数据流程图

在实际的生产环境，电商产生的实时数据需要更多的统计如热门商品统计、实时热门页面流量统计、实时访问流量统计、APP市场推广统计，页面广告点击量统计、页面广告黑名单过滤、恶意登录监控、订单支付失效监控、支付实时对账数据全部都需要做实时流数据计算，对热门商品都需要优先计算，数据流量如图8-7所示。

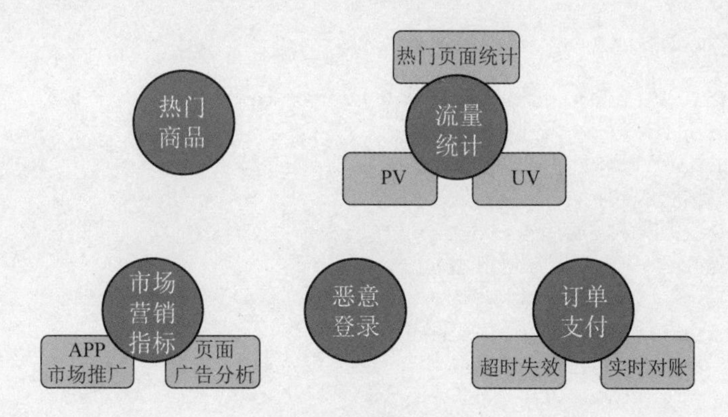

图8-7 数据流量图

8.4.2 MySQL配置文件

在resource源码包下创建comerce.peoperties数据链接，配置数据源连字符串，实现代码如下所示：

```
jdbc.url=jdbc:mysql://192.168.13.100:3306/course_learn?useUnicode=
true&characterEncoding=utf8&serverTimezone=Asia/Shanghai&useSSL=false
jdbc.user=root
jdbc.password=000000
```

8.4.3 创建读取配置文件的工具类

在com.cqsx.qzpoint.util中创建ConfigurationManager类，代码如下所示：

```
package com.cqsx.qzpoint.util;
```

```
import java.io.InputStream;
import java.util.Properties;

/**
 *
 * 读取配置文件工具类
 */
public class ConfigurationManager {

  private static Properties prop = new Properties();

  static {
    try {
      InputStream inputStream = ConfigurationManager.class.getClassLoader()
          .getResourceAsStream("comerce.properties");
      prop.load(inputStream);
    } catch (Exception e) {
      e.printStackTrace();
    }
  }
  //获取配置项
  public static String getProperty(String key) {
    return prop.getProperty(key);
  }
  //获取布尔类型的配置项
  public static boolean getBoolean(String key) {
    String value = prop.getProperty(key);
    try {
      return Boolean.valueOf(value);
    } catch (Exception e) {
      e.printStackTrace();
    }
    return false;
  }

}
```

8.4.4 Json解析工具类

在com.cqsx.qz.point.util中创建ParseJsonData类，代码如下所示；

```
package com.cqsx.qzpoint.util;

import com.alibaba.fastjson.JSONObject;


public class ParseJsonData {

    public static JSONObject getJsonData(String data) {
        try {
            return JSONObject.parseObject(data);
```

```
        } catch (Exception e) {
            return null;
        }
    }
}
```

8.4.5 创建Druid连接池

在com.atgugiu.qzpoint.util中创建DataSourceUtil类，建立数据连接，初始化大小，最大连接，最小连接，等待时长。配置多久进行一次检测，检测需要关闭的连接单位为毫秒，配置连接在连接池中的最小生存时间单位为毫秒，配置连接在连接池中的最小生存时间单位为毫秒。代码如下所示：

```
package com.cqsx.qzpoint.util;
import com.alibaba.druid.pool.DruidDataSourceFactory;
import javax.sql.DataSource;
import java.io.Serializable;
import java.sql.*;
import java.util.Properties;
/**
 *
 */
public class DataSourceUtil implements Serializable {
    public static DataSource dataSource = null;
    static {
        try {
            Properties props = new Properties();
            props.setProperty("url", ConfigurationManager.
            getProperty("jdbc.url"));
            props.setProperty("username", ConfigurationManager.
            getProperty("jdbc.user"));
            props.setProperty("password", ConfigurationManager.
            getProperty("jdbc.password"));
            props.setProperty("initialSize","5");
            props.setProperty("maxActive","10");
            props.setProperty("minIdle","5");
            props.setProperty("maxWait","60000");
            props.setProperty("timeBetweenEvictionRunsMillis","2000");
            //
            props.setProperty("minEvictableIdleTimeMillis","600000");
            //
            props.setProperty("maxEvictableIdleTimeMillis","900000");
            //配置连接在连接池中最大生存时间单位毫秒
            props.setProperty("validationQuery","select 1");
            props.setProperty("testWhileIdle","true");
            props.setProperty("testOnBorrow","false");
            props.setProperty("testOnReturn","false");
            props.setProperty("keepAlive","true");
            props.setProperty("phyMaxUseCount","100000");
            props.setProperty("driverClassName","com.mysql.jdbc.Driver");
```

```
            dataSource = DruidDataSourceFactory.createDataSource(props);
        } catch (Exception e) {
            e.printStackTrace();
        }
    }

    //public static Connection getConnection() throws SQLException {
        return dataSource.getConnection();
    }
    public static void closeResource(ResultSet resultSet,
        PreparedStatement preparedStatement,
                                         Connection connection) {
        //关闭结果集
        //ctrl+alt+m将java语句抽取成方法
        closeResultSet(resultSet);
        //关闭语句执行者
        closePrepareStatement(preparedStatement);
        //关闭连接
        closeConnection(connection);
    }

    private static void closeConnection(Connection connection) {
        if (connection != null) {
            try {
                connection.close();
            } catch (SQLException e) {
                e.printStackTrace();
            }
        }
    }

    private static void closePrepareStatement(PreparedStatement
        preparedStatement) {
        if (preparedStatement != null) {
            try {
                preparedStatement.close();
            } catch (SQLException e) {
                e.printStackTrace();
            }
        }
    }

    private static void closeResultSet(ResultSet resultSet) {
        if (resultSet != null) {
            try {
                resultSet.close();
            } catch (SQLException e) {
                e.printStackTrace();
            }
        }
    }
}
```

8.4.6 创建MySQL的代理类

在com.cqsx.qzpoint.util中创建SqlProxy类，执行查询数据，在查询数据中定义DataSourceUtil，实现代码如下所示：

```
package com.cqsx.qzpoint.util

import java.sql.{Connection,PreparedStatement,ResultSet}

trait QueryCallback {
  def process(rs:ResultSet)
}

class SqlProxy {
  private var rs:ResultSet = _
  private var psmt:PreparedStatement = _

  /**
    *
    * @param conn
    * @param sql
    * @param params
    * @return
    */
  def executeUpdate(conn:Connection,sql:String,params:Array[Any]):Int = {
    var rtn = 0
    try {
      psmt = conn.prepareStatement(sql)
      if (params != null && params.length > 0) {
        for (i <- 0 until params.length) {
          psmt.setObject(i + 1,params(i))
        }
      }
      rtn = psmt.executeUpdate()
    } catch {
      case e:Exception => e.printStackTrace()
    }
    rtn
  }

  /**
    *
    * @param conn
    * @param sql
    * @param params
    * @return
    */
  def executeQuery(conn:Connection,sql:String,params:Array[Any],
  queryCallback:QueryCallback) = {
    rs = null
    try {
      psmt = conn.prepareStatement(sql)
```

```
      if (params != null && params.length > 0) {
        for (i <- 0 until params.length) {
          psmt.setObject(i + 1,params(i))
        }
      }
      rs = psmt.executeQuery()
      queryCallback.process(rs)
    } catch {
      case e: Exception => e.printStackTrace()
    }
  }

  def shutdown(conn: Connection):Unit = DataSourceUtil.
closeResource(rs,psmt,conn)
}
```

8.4.7 访问人数统计

通过查询MySQL的访问记录来判断数据是否有偏移量，再次判断本地是否有偏移量。根据偏移量计算出访问和消费的人数，如果无偏移量，则重新计算访问和消费的人数。实现代码如下所示：

```
package com.cqsx.qzpoint.streaming
import java.lang
import java.sql.ResultSet
import java.util.Random
import com.cqsx.qzpoint.util.{DataSourceUtil,QueryCallback,SqlProxy}
import org.apache.kafka.clients.consumer.ConsumerRecord
import org.apache.kafka.common.TopicPartition
import org.apache.kafka.common.serialization.StringDeserializer
import org.apache.spark.SparkConf
import org.apache.spark.streaming.dstream.InputDStream
import org.apache.spark.streaming.kafka010._
import org.apache.spark.streaming.{Seconds,StreamingContext}
import scala.collection.mutable
object RegisterStreaming {
  private val groupid = "register_group_test"
  def main(args:Array[String]):Unit = {
    val conf = new SparkConf().setAppName(this.getClass.getSimpleName).
      setMaster("local[2]")
      .set("spark.streaming.kafka.maxRatePerPartition","50")
      .set("spark.streaming.stopGracefullyOnShutdown","true")
    val ssc = new StreamingContext(conf,Seconds(3))
    val topics = Array("register_topic")
    val kafkaMap:Map[String,Object] = Map[String,Object](
      "bootstrap.servers" -> "server1:9092,hadoop103:9092,
       hadoop104:9092",
      "key.deserializer" -> classOf[StringDeserializer],
      "value.deserializer" -> classOf[StringDeserializer],
```

```
    "group.id" -> groupid,
    "auto.offset.reset" -> "earliest",
    "enable.auto.commit" -> (false:lang.Boolean)
  )
  ssc.checkpoint("hdfs://server1:9000/user/cqsx/sparkstreaming/
      checkpoint")

  val sqlProxy = new SqlProxy()
  val offsetMap = new mutable.HashMap[TopicPartition,Long]()
  val client = DataSourceUtil.getConnection
  try {
    sqlProxy.executeQuery(client,"select * from `offset_manager`
      where groupid=?",Array(groupid),new QueryCallback {
      override def process(rs:ResultSet):Unit = {
        while (rs.next()) {
          val model = new TopicPartition(rs.getString(2),rs.getInt(3))
          val offset = rs.getLong(4)
          offsetMap.put(model,offset)
        }
        rs.close()
      }
    })
  } catch {
    case e:Exception => e.printStackTrace()
  } finally {
    sqlProxy.shutdown(client)
  }
  val stream:InputDStream[ConsumerRecord[String,String]] =
    if(offsetMap.isEmpty) {
    KafkaUtils.createDirectStream(
      ssc,LocationStrategies.PreferConsistent,ConsumerStrategies.
        Subscribe[String,String](topics,kafkaMap))
  } else {
    KafkaUtils.createDirectStream(
      ssc,LocationStrategies.PreferConsistent,ConsumerStrategies.
        Subscribe[String,String](topics,kafkaMap,offsetMap))
  }
  val resultDStream = stream.filter(item =>
    item.value().split("\t").length == 3).
    mapPartitions(partitions => {
      partitions.map(item => {
        val line = item.value()
        val arr = line.split("\t")
        val app_name = arr(1) match {
          case "1" => "PC"
          case "2" => "APP"
          case _ => "Other"
        }
        (app_name,1)
      })
    })
  resultDStream.cache()
```

```
  resultDStream.reduceByKeyAndWindow((x:Int,y:Int) => x + y,
     Seconds(60),Seconds(6)).print()
   val updateFunc = (values:Seq[Int],state:Option[Int]) => {
     val currentCount = values.sum
     val previousCount = state.getOrElse(0)
     Some(currentCount + previousCount)
   }
   resultDStream.updateStateByKey(updateFunc).print()
       val dsStream = stream.filter(item =>
         item.value().split("\t").length == 3)
         .mapPartitions(partitions =>
           partitions.map(item => {
             val rand = new Random()
             val line = item.value()
             val arr = line.split("\t")
             val app_id = arr(1)
             (rand.nextInt(3) + "_" + app_id,1)
         }))
      dsStream.print()
       val a = dsStream.reduceByKey(_ + _)
       a.print()
       a.map(item => {
         val appid = item._1.split("_")(1)
         (appid,item._2)
      }).reduceByKey(_ + _).print()
   stream.foreachRDD(rdd => {
     val sqlProxy = new SqlProxy()
     val client = DataSourceUtil.getConnection
     try {
       val offsetRanges:Array[OffsetRange] = rdd.asInstanceOf
         [HasOffsetRanges].offsetRanges
       for (or <- offsetRanges) {
         sqlProxy.executeUpdate(client,"replace into `offset_manager`
           (groupid,topic,`partition`,untilOffset) values(?,?,?,?)",
           Array(groupid,or.topic,or.partition.toString,or.untilOffset))
       }
     } catch {
       case e:Exception => e.printStackTrace()
     } finally {
       sqlProxy.shutdown(client)
     }
   })
   ssc.start()
   ssc.awaitTermination()
 }
}
```

8.4.8 实时统计

实现实时采集统计时，程序员会调用Kafka基础接口包，通过Flink读取Kafka生成的数据，使用Kafka创建电商消息数据队列，并对生产的信息进行汇总。分别调用Spark实时数据流完成实时数的计算字符串。实现对热度数据的流处理能力，实现代码如下所示：

```
package com.cqsx.qzpoint.streaming
import java.lang
import java.sql.{Connection,ResultSet}
import java.time.LocalDateTime
import java.time.format.DateTimeFormatter
import com.cqsx.qzpoint.util.{DataSourceUtil,QueryCallback,SqlProxy}
import org.apache.kafka.clients.consumer.ConsumerRecord
import org.apache.kafka.common.TopicPartition
import org.apache.kafka.common.serialization.StringDeserializer
import org.apache.spark.SparkConf
import org.apache.spark.streaming.dstream.InputDStream
import org.apache.spark.streaming.kafka010._
import org.apache.spark.streaming.{Seconds,StreamingContext}
import scala.collection.mutable
/**
  * 热度统计
  */
object QzPointStreaming {

  private val groupid = "qz_point_group"

  def main(args: Array[String]):Unit = {
    val conf = new SparkConf().setAppName(this.getClass.getSimpleName).
      setMaster("local[*]")
      .set("spark.streaming.kafka.maxRatePerPartition","50")
      .set("spark.streaming.stopGracefullyOnShutdown","true")
    val ssc = new StreamingContext(conf, Seconds(3))
    val topics = Array("qz_log")
    val kafkaMap:Map[String,Object] = Map[String,Object](
      "bootstrap.servers" -> "server1:9092,hadoop103:9092,
       hadoop104:9092",
      "key.deserializer" -> classOf[StringDeserializer],
      "value.deserializer" -> classOf[StringDeserializer],
      "group.id" -> groupid,
      "auto.offset.reset" -> "earliest",
      "enable.auto.commit" -> (false:lang.Boolean)
    )

    val sqlProxy = new SqlProxy()
    val offsetMap = new mutable.HashMap[TopicPartition,Long]()
    val client = DataSourceUtil.getConnection
    try {
      sqlProxy.executeQuery(client,"select * from `offset_manager`
        where groupid=?",Array(groupid),new QueryCallback {
        override def process(rs:ResultSet):Unit = {
          while (rs.next()) {
            val model = new TopicPartition(rs.getString(2),rs.getInt(3))
            val offset = rs.getLong(4)
            offsetMap.put(model,offset)
          }
```

```
        rs.close()
      }
    })
  } catch {
    case e: Exception => e.printStackTrace()
  } finally {
    sqlProxy.shutdown(client)
  }
  val stream:InputDStream[ConsumerRecord[String,String]] =
    if(offsetMap.isEmpty) {
    KafkaUtils.createDirectStream(
      ssc,LocationStrategies.PreferConsistent,ConsumerStrategies.
        Subscribe[String,String](topics,kafkaMap))
  } else {
    KafkaUtils.createDirectStream(
      ssc,LocationStrategies.PreferConsistent,ConsumerStrategies.
        Subscribe[String,String](topics,kafkaMap,offsetMap))
  }

  val dsStream = stream.filter(item =>
    item.value().split("\t").length == 6).
    mapPartitions(partition => partition.map(item => {
      val line = item.value()
      val arr = line.split("\t")
      val uid = arr(0) //用户id
      val courseid = arr(1)
      val pointid = arr(2)
      val questionid = arr(3) val istrue =
        arr(4)          val createtime = arr(5)
      (uid, courseid,pointid,questionid,istrue,createtime)
    }))
  dsStream.foreachRDD(rdd => {

    val groupRdd = rdd.groupBy(item => item._1 + "-" + item._2 +
      "-" + item._3)
    groupRdd.foreachPartition(partition => {
          val sqlProxy = new SqlProxy()
      val client = DataSourceUtil.getConnection
      try {
        partition.foreach { case (key,iters) =>
          qzQuestionUpdate(key,iters,sqlProxy,client)
        }
      } catch {
        case e:Exception => e.printStackTrace()
      }
      finally {
        sqlProxy.shutdown(client)
      }
    }
    )
  })
```

```
  stream.foreachRDD(rdd => {
    val sqlProxy = new SqlProxy()
    val client = DataSourceUtil.getConnection
    try {
      val offsetRanges:Array[OffsetRange] = rdd.asInstanceOf
        [HasOffsetRanges].offsetRanges
      for (or <- offsetRanges) {
        sqlProxy.executeUpdate(client,"replace into `offset_manager`
          (groupid,topic,`partition`,untilOffset) values(?,?,?,?)",
          Array(groupid,or.topic,or.partition.toString,or.untilOffset))
      }
    } catch {
      case e: Exception => e.printStackTrace()
    } finally {
      sqlProxy.shutdown(client)
    }
  })
  ssc.start()
  ssc.awaitTermination()
}

/**
  *
  *
  * @param key
  * @param iters
  * @param sqlProxy
  * @param client
  * @return
  */
def qzQuestionUpdate(key: String,iters:Iterable[(String,String,String,
  String,String,String)],sqlProxy:SqlProxy,client:Connection) = {
  val keys = key.split("-")
  val userid = keys(0).toInt
  val courseid = keys(1).toInt
  val pointid = keys(2).toInt
  val array = iters.toArray
  val questionids = array.map(_._4).distinct
   var questionids_history: Array[String] = Array()
  sqlProxy.executeQuery(client, "select questionids from qz_point_
    history where userid=? and courseid=? and pointid=?",
    Array(userid,courseid,pointid),new QueryCallback {
      override def process(rs:ResultSet):Unit = {
        while (rs.next()) {
          questionids_history = rs.getString(1).split(",")
        }
        rs.close()          }
    })

  val resultQuestionid = questionids.union(questionids_history).distinct
  val countSize = resultQuestionid.length
  val resultQuestionid_str = resultQuestionid.mkString(",")
```

```
    val qz_count = questionids.length     var qz_sum = array.length
    var qz_istrue = array.filter(_._5.equals("1")).size
    val createtime = array.map(_._6).min
    val updatetime = DateTimeFormatter.ofPattern("yyyy-MM-dd HH:
      mm:ss").format(LocalDateTime.now())
    sqlProxy.executeUpdate(client, "insert into qz_point_history(userid,
      courseid,pointid,questionids,createtime,updatetime) values(?,?,?,?,?,?)
      " +" on duplicate key update questionids=?,updatetime=?", Array
      (userid, courseid, pointid, resultQuestionid_str, createtime,
      createtime, resultQuestionid_str, updatetime))
  var qzSum_history = 0
    var istrue_history = 0
    sqlProxy.executeQuery(client, "select qz_sum,qz_istrue from
      qz_point_detail where userid=? and courseid=? and pointid=?",
      Array(userid, courseid, pointid), new QueryCallback {
        override def process(rs: ResultSet): Unit = {
          while (rs.next()) {
            qzSum_history += rs.getInt(1)
            istrue_history += rs.getInt(2)
          }
          rs.close()
        }
      })
    qz_sum += qzSum_history
    qz_istrue += istrue_history
    val correct_rate = qz_istrue.toDouble / qz_sum.toDouble
    val qz_detail_rate = countSize.toDouble / 30
    val mastery_rate = qz_detail_rate * correct_rate
    sqlProxy.executeUpdate(client, "insert into qz_point_detail(userid,
      courseid,pointid,qz_sum,qz_count,qz_istrue,correct_rate,mastery_
      rate,createtime,updatetime)" +
      " values(?,?,?,?,?,?,?,?,?,?) on duplicate key update qz_sum=?,
      qz_count=?,qz_istrue=?,correct_rate=?,mastery_rate=?,
      updatetime=?",Array(userid, courseid, pointid, qz_sum, countSize,
      qz_istrue, correct_rate, mastery_rate, createtime, updatetime,
      qz_sum, countSize, qz_istrue, correct_rate, mastery_rate,
      updatetime))

  }
}
```

8.4.9 实时统计商品

实现一个“实时热门商品”的需求，我们可以将“实时热门商品”写作程序员更好理解的需求：每隔1分钟输出最近一小时内点击量最多的前N个商品。取出业务时间戳，告诉Flink框架基于业务时间做窗口，过滤出点击行为数据，按一小时的窗口大小，每1分钟统计一次，做滑动窗口聚合Sliding Window一份访问用户行为数据集。本数据集包含了第三方平台上某一天随机一百万用户的所有行为，包括点击、购买、加购、收藏。数据集的组织形式与MovieLens-20M类似，即数据集的每一行表示一个用户行为，由用户ID、商品ID、商品类目ID、行为类型和时间戳组成，并以逗号分隔。关于数据集中每一列的详细代码如下：

```
package com.cqsx.qzpoint.streaming
import java.lang
import java.sql.{Connection, ResultSet}
import java.text.NumberFormat
import com.cqsx.qzpoint.util.{DataSourceUtil, ParseJsonData,
QueryCallback, SqlProxy}
import org.apache.kafka.clients.consumer.ConsumerRecord
import org.apache.kafka.common.TopicPartition
import org.apache.kafka.common.serialization.StringDeserializer
import org.apache.spark.streaming.dstream.InputDStream
import org.apache.spark.streaming.kafka010._
import org.apache.spark.streaming.{Seconds, StreamingContext}
import org.apache.spark.{SparkConf, SparkFiles}
import org.lionsoul.ip2region.{DbConfig, DbSearcher}
import scala.collection.mutable
import scala.collection.mutable.ArrayBuffer
/**
  * 页面实时计算
  */
object PageStreaming {
  private val groupid = "vip_count_groupid"

  def main(args: Array[String]): Unit = {
    val conf = new SparkConf().setAppName(this.getClass.getSimpleName).
      setMaster("local[*]")
      .set("spark.streaming.kafka.maxRatePerPartition", "30")
      .set("spark.streaming.stopGracefullyOnShutdown", "true")

    val ssc = new StreamingContext(conf, Seconds(3))
    val topics = Array("page_topic")
    val kafkaMap: Map[String, Object] = Map[String, Object](
      "bootstrap.servers" -> "server1:9092,hadoop103:9092,
       hadoop104:9092",
      "key.deserializer" -> classOf[StringDeserializer],
      "value.deserializer" -> classOf[StringDeserializer],
      "group.id" -> groupid,
      "auto.offset.reset" -> "earliest",
      "enable.auto.commit" -> (false: lang.Boolean)
    )

    val sqlProxy = new SqlProxy()
    val offsetMap = new mutable.HashMap[TopicPartition, Long]()
    val client = DataSourceUtil.getConnection
    try {
      sqlProxy.executeQuery(client, "select *from `offset_manager`
        where groupid=?", Array(groupid), new QueryCallback {
        override def process(rs: ResultSet): Unit = {
          while (rs.next()) {
            val model = new TopicPartition(rs.getString(2), rs.getInt(3))
            val offset = rs.getLong(4)
```

```
          offsetMap.put(model, offset)
        }
        rs.close()
      }
    })
  } catch {
    case e: Exception => e.printStackTrace()
  } finally {
    sqlProxy.shutdown(client)
  }

  val stream: InputDStream[ConsumerRecord[String, String]] =
    if(offsetMap.isEmpty) {
    KafkaUtils.createDirectStream(
      ssc, LocationStrategies.PreferConsistent, ConsumerStrategies.
        Subscribe[String, String](topics, kafkaMap))
  } else {
    KafkaUtils.createDirectStream(
      ssc, LocationStrategies.PreferConsistent, ConsumerStrategies.
        Subscribe[String, String](topics, kafkaMap, offsetMap))
  }

  val dsStream = stream.map(item => item.value()).mapPartitions
    (partition => {
    partition.map(item => {
      val jsonObject = ParseJsonData.getJsonData(item)
      val uid = if (jsonObject.containsKey("uid")) jsonObject.
        getString("uid") else ""
      val app_id = if (jsonObject.containsKey("app_id")) jsonObject.
        getString("app_id") else ""
      val device_id = if (jsonObject.containsKey("device_id"))
        jsonObject.getString("device_id") else ""
      val ip = if (jsonObject.containsKey("ip")) jsonObject.
        getString("ip") else ""
      val last_page_id = if (jsonObject.containsKey("last_page_id"))
        jsonObject.getString("last_page_id") else ""
      val pageid = if (jsonObject.containsKey("page_id")) jsonObject.
        getString("page_id") else ""
      val next_page_id = if (jsonObject.containsKey("next_page_id"))
        jsonObject.getString("next_page_id") else ""
      (uid, app_id, device_id, ip, last_page_id, pageid, next_page_id)
    })
  }).filter(item => {
    !item._5.equals("") && !item._6.equals("") && !item._7.equals("")
  })
  dsStream.cache()
  val pageValueDStream = dsStream.map(item => (item._5 + "_" +
    item._6 + "_" + item._7, 1))
  val resultDStream = pageValueDStream.reduceByKey(_ + _)
  resultDStream.foreachRDD(rdd => {
    rdd.foreachPartition(partition => {
```

```
    val sqlProxy = new SqlProxy()
    val client = DataSourceUtil.getConnection
    try {
      partition.foreach(item => {
        calcPageJumpCount(sqlProxy, item, client)
      })
    } catch {
      case e: Exception => e.printStackTrace()
    } finally {
      sqlProxy.shutdown(client)
    }
  })
})
ssc.sparkContext.addFile(this.getClass.getResource(
  "/ip2region.db").getPath)
val ipDStream = dsStream.mapPartitions(patitions => {
  val dbFile = SparkFiles.get("ip2region.db")
  val ipsearch = new DbSearcher(new DbConfig(), dbFile)
  patitions.map { item =>
    val ip = item._4
    val province = ipsearch.memorySearch(ip).getRegion().
      split("\\|")(2)          (province, 1l)
  }
}).reduceByKey(_ + _)
ipDStream.foreachRDD(rdd => {

  val ipSqlProxy = new SqlProxy()
  val ipClient = DataSourceUtil.getConnection
  try {
    val history_data = new ArrayBuffer[(String, Long)]()
    ipSqlProxy.executeQuery(ipClient, "select province,num from
      tmp_city_num_detail", null, new QueryCallback {
      override def process(rs: ResultSet): Unit = {
        while (rs.next()) {
          val tuple = (rs.getString(1), rs.getLong(2))
          history_data += tuple
        }
      }
    })
    val history_rdd = ssc.sparkContext.makeRDD(history_data)
    val resultRdd = history_rdd.fullOuterJoin(rdd).map(item => {
      val province = item._1
      val nums = item._2._1.getOrElse(0l) + item._2._2.getOrElse(0l)
      (province, nums)
    })
    resultRdd.foreachPartition(partitions => {
      val sqlProxy = new SqlProxy()
      val client = DataSourceUtil.getConnection
      try {
        partitions.foreach(item => {
          val province = item._1
          val num = item._2
```

```
            sqlProxy.executeUpdate(client, "insert into tmp_city_num_
              detail(province,num)values(?,?) on duplicate key update
              num=?",Array(province, num, num))
          })
        } catch {
          case e: Exception => e.printStackTrace()
        } finally {
          sqlProxy.shutdown(client)
        }
      })
      val top3Rdd = resultRdd.sortBy[Long](_._2, false).take(3)
      sqlProxy.executeUpdate(ipClient, "truncate table top_city_
        num", null)
      top3Rdd.foreach(item => {
        sqlProxy.executeUpdate(ipClient, "insert into top_city_num
          (province,num) values(?,?)", Array(item._1, item._2))
      })
    } catch {
      case e: Exception => e.printStackTrace()
    } finally {
      sqlProxy.shutdown(ipClient)
    }
  })

  stream.foreachRDD(rdd => {
    val sqlProxy = new SqlProxy()
    val client = DataSourceUtil.getConnection
    try {
      calcJumRate(sqlProxy, client)
      val offsetRanges: Array[OffsetRange] = rdd.asInstanceOf
        [HasOffsetRanges].offsetRanges
      for (or <- offsetRanges) {
        sqlProxy.executeUpdate(client, "replace into `offset_manager`
          (groupid,topic,`partition`,untilOffset) values(?,?,?,?)",
          Array(groupid, or.topic, or.partition.toString, or.untilOffset))
      }
    } catch {
      case e: Exception => e.printStackTrace()
    } finally {
      sqlProxy.shutdown(client)
    }
  })
  ssc.start()
  ssc.awaitTermination()
}
/**
  * 计算Web
  *
  * @param sqlProxy
  * @param item
  * @param client
  */
```

```
def calcPageJumpCount(sqlProxy: SqlProxy, item: (String, Int),
  client: Connection): Unit = {
  val keys = item._1.split("_")
  var num: Long = item._2
  val page_id = keys(1).toInt
  val last_page_id = keys(0).toInt
  val next_page_id = keys(2).toInt

  sqlProxy.executeQuery(client, "select num from page_jump_rate where
    page_id=?", Array(page_id), new QueryCallback {
    override def process(rs: ResultSet): Unit = {
      while (rs.next()) {
        num += rs.getLong(1)
      }
      rs.close()
    }

    if (page_id == 1) {
      sqlProxy.executeUpdate(client, "insert into page_jump_rate
        (last_page_id,page_id,next_page_id,num,jump_rate)" +
        "values(?,?,?,?,?) on duplicate key update num=num+?",
        Array(last_page_id, page_id, next_page_id, num, "100%", num))
    } else {
      sqlProxy.executeUpdate(client, "insert into page_jump_
        rate(last_page_id,page_id,next_page_id,num)" +
        "values(?,?,?,?) on duplicate key update num=num+?",
        Array(last_page_id, page_id, next_page_id, num, num))
    }
  })
}
/**
  * 计算业务数据
  */
def calcJumRate(sqlProxy: SqlProxy, client: Connection): Unit = {
  var page1_num = 0l
  var page2_num = 0l
  var page3_num = 0l
  sqlProxy.executeQuery(client, "select num from page_jump_rate where
    page_id=?", Array(1), new QueryCallback {
    override def process(rs: ResultSet): Unit = {
      while (rs.next()) {
        page1_num = rs.getLong(1)
      }
    }
  })
  sqlProxy.executeQuery(client, "select num from page_jump_rate where
    page_id=?", Array(2), new QueryCallback {
    override def process(rs: ResultSet): Unit = {
      while (rs.next()) {
        page2_num = rs.getLong(1)
      }
    }
```

```
    })
    sqlProxy.executeQuery(client, "select num from page_jump_rate where
      page_id=?", Array(3), new QueryCallback {
      override def process(rs: ResultSet): Unit = {
        while (rs.next()) {
          page3_num = rs.getLong(1)
        }
      }
    })
    val nf = NumberFormat.getPercentInstance
    val page1ToPage2Rate = if (page1_num == 0) "0%" else nf.
      format(page2_num.toDouble / page1_num.toDouble)
    val page2ToPage3Rate = if (page2_num == 0) "0%" else nf.
      format(page3_num.toDouble / page2_num.toDouble)
    sqlProxy.executeUpdate(client, "update page_jump_rate set
      jump_rate=? where page_id=?", Array(page1ToPage2Rate, 2))
    sqlProxy.executeUpdate(client, "update page_jump_rate set
      jump_rate=? where page_id=?", Array(page2ToPage3Rate, 3))
  }
}
```

8.4.10 实时数据统计

智慧农业电商平台要实现的模块是“实时流量统计”。对于一个智慧农业电商平台而言，用户登录的入口流量、不同页面的访问流量都是值得分析的重要数据，而这些数据可以简单地从web服务器的日志中提取出来。我们在这里先实现“热门页面浏览数”的统计，也就是读取服务器日志中的每一行log，统计在一段时间内用户访问每一个URL的次数，然后排序输出显示。具体做法为：每隔5秒输出最近10分钟内访问量最多的前N个URL。可以看出，这个需求与之前“实时热门商品统计”非常类似，所以完全可以借鉴此前的代码。实现实时数据统计程序如以下代码所示。

```
package com.cqsx.qzpoint.streaming
import java.lang
import java.sql.{Connection,ResultSet}
import com.cqsx.qzpoint.bean.LearnModel
import com.cqsx.qzpoint.util.{DataSourceUtil,ParseJsonData,
QueryCallback,SqlProxy}
import org.apache.kafka.clients.consumer.ConsumerRecord
import org.apache.kafka.common.TopicPartition
import org.apache.kafka.common.serialization.StringDeserializer
import org.apache.spark.SparkConf
import org.apache.spark.streaming.dstream.InputDStream
import org.apache.spark.streaming.kafka010._
import org.apache.spark.streaming.{Seconds,StreamingContext}
import scala.collection.mutable
import scala.collection.mutable.ArrayBuffer
object CourseLearnStreaming {
    private val groupid = "course_learn_test1"
    def main(args: Array[String]):Unit = {
        val conf = new SparkConf().setAppName(this.getClass.getSimpleName)
```

```
        .set("spark.streaming.kafka.maxRatePerPartition","30")
        .set("spark.streaming.stopGracefullyOnShutdown","true")

    val ssc = new StreamingContext(conf,Seconds(3))
    val topics = Array("course_learn")
    val kafkaMap:Map[String,Object] = Map[String,Object](
        "bootstrap.servers" -> "server1:9092,hadoop103:9092,
         hadoop104:9092",
        "key.deserializer" -> classOf[StringDeserializer],
        "value.deserializer" -> classOf[StringDeserializer],
        "group.id" -> groupid,
        "auto.offset.reset" -> "earliest",
        "enable.auto.commit" -> (false:lang.Boolean)
    )
    val sqlProxy = new SqlProxy()
    val offsetMap = new mutable.HashMap[TopicPartition,Long]()
    val client = DataSourceUtil.getConnection
try {
    sqlProxy.executeQuery(client,"select *from `offset_manager`
        where groupid=?",Array(groupid),new QueryCallback {
            override def process(rs:ResultSet):Unit = {
                while (rs.next()) {
                    val model = new TopicPartition(
                      rs.getString(2),rs.getInt(3))
                    val offset = rs.getLong(4)
                    offsetMap.put(model, offset)
                }
            rs.close()
        }
    })
} catch {
    case e:Exception => e.printStackTrace()
} finally {
    sqlProxy.shutdown(client)
}
val stream:InputDStream[ConsumerRecord[String,String]] =
  if(offsetMap.isEmpty) {
    KafkaUtils.createDirectStream(
        ssc,LocationStrategies.PreferConsistent,
        ConsumerStrategies.Subscribe[String,String](topics,kafkaMap))
    } else {
        KafkaUtils.createDirectStream(
            ssc,LocationStrategies.PreferConsistent,
              ConsumerStrategies.Subscribe[
                String,String](topics,kafkaMap,offsetMap))
    }
val dsStream = stream.mapPartitions(partitions => {
    partitions.map(item => {
        val json = item.value()
        val jsonObject = ParseJsonData.getJsonData(json)
        val userId = jsonObject.getIntValue("uid")
        val cwareid = jsonObject.getIntValue("cwareid")
```

```
        val videoId = jsonObject.getIntValue("videoid")
        val chapterId = jsonObject.getIntValue("chapterid")
        val edutypeId = jsonObject.getIntValue("edutypeid")
        val subjectId = jsonObject.getIntValue("subjectid")
        val sourceType = jsonObject.getString("sourceType")
        val speed = jsonObject.getIntValue("speed")
        val ts = jsonObject.getLong("ts")
        val te = jsonObject.getLong("te")
        val ps = jsonObject.getIntValue("ps")
        val pe = jsonObject.getIntValue("pe")
        LearnModel(userId,cwareid,videoId,chapterId,edutypeId,
        subjectId,sourceType,speed,ts,te,ps,pe)
    })
})
dsStream.foreachRDD(rdd => {
    rdd.cache()
    rdd.groupBy(item => item.userId + "_" + item.cwareId +
        "_" + item.videoId).foreachPartition(partitoins => {
        val sqlProxy = new SqlProxy()
        val client = DataSourceUtil.getConnection
        try {
            partitoins.foreach { case (key,iters) =>
                calcVideoTime(key,iters,sqlProxy,client)
            }
        } catch {
            case e: Exception => e.printStackTrace()
        } finally {
            sqlProxy.shutdown(client)
        }
    })
        rdd.mapPartitions(partitions => {
            partitions.map(item => {
                val totaltime = Math.ceil((item.te - item.ts) /
                  1000).toLong
                val key = item.chapterId
                (key,totaltime)
            })
        }).reduceByKey(_ + _)
            .foreachPartition(partitoins => {
                val sqlProxy = new SqlProxy()
                val client = DataSourceUtil.getConnection
                try {
                    partitoins.foreach(item => {
                        sqlProxy.executeUpdate(client,
                        "insert into chapter_learn_detail(chapterid,
                         totaltime) values(?,?) on duplicate key" +
                        "update totaltime=totaltime+?",
                         Array(item._1, item._2, item._2))
                    })
      } catch {
        case e: Exception => e.printStackTrace()
      } finally {
```

```
        sqlProxy.shutdown(client)
      }
    })

  rdd.mapPartitions(partitions => {
    partitions.map(item => {
      val totaltime = Math.ceil((item.te - item.ts) / 1000).toLong
      val key = item.cwareId
      (key, totaltime)
    })
  }).reduceByKey(_ + _).foreachPartition(partitions => {
    val sqlProxy = new SqlProxy()
    val client = DataSourceUtil.getConnection
    try {
      partitions.foreach(item => {
        sqlProxy.executeUpdate(client, "insert into cwareid_learn_
          detail(cwareid,totaltime) values(?,?) on duplicate key "
          +"update totaltime=totaltime+?", Array(item._1, item._2,
          item._2))
      })
    } catch {
      case e: Exception => e.printStackTrace()
    } finally {
      sqlProxy.shutdown(client)
    }
  })
  rdd.mapPartitions(partitions => {
    partitions.map(item => {
      val totaltime = Math.ceil((item.te - item.ts) / 1000).toLong
      val key = item.edutypeId
      (key, totaltime)
    })
  }).reduceByKey(_ + _).foreachPartition(partitions => {
    val sqlProxy = new SqlProxy()
    val client = DataSourceUtil.getConnection
    try {
      partitions.foreach(item => {
        sqlProxy.executeUpdate(client, "insert into edutype_learn_
          detail(edutypeid,totaltime) values(?,?) on duplicate key " +
          "update totaltime=totaltime+?", Array(item._1, item._2,
          item._2))
      })
    } catch {
      case e: Exception => e.printStackTrace()
    } finally {
      sqlProxy.shutdown(client)
    }
  })
  rdd.mapPartitions(partitions => {
    partitions.map(item => {
      val totaltime = Math.ceil((item.te - item.ts) / 1000).toLong
      val key = item.sourceType
```

```
        (key, totaltime)
      })
    }).reduceByKey(_ + _).foreachPartition(partitions => {
      val sqlProxy = new SqlProxy()
      val client = DataSourceUtil.getConnection
      try {
        partitions.foreach(item => {
          sqlProxy.executeUpdate(client, "insert into sourcetype_
            learn_detail (sourcetype_learn,totaltime) values(?,?)
            on duplicate key " + "update totaltime=totaltime+?",
            Array(item._1, item._2, item._2))
        })
      } catch {
        case e: Exception => e.printStackTrace()
      } finally {
        sqlProxy.shutdown(client)
      }
    })
    rdd.mapPartitions(partitions => {
      partitions.map(item => {
        val totaltime = Math.ceil((item.te - item.ts) / 1000).toLong
        val key = item.subjectId
        (key, totaltime)
      })
    }).reduceByKey(_ + _).foreachPartition(partitons => {
      val sqlProxy = new SqlProxy()
      val clinet = DataSourceUtil.getConnection
      try {
        partitons.foreach(item => {
          sqlProxy.executeUpdate(clinet, "insert into subject_learn_
            detail(subjectid,totaltime) values(?,?) on duplicate
            key " +"update totaltime=totaltime+?", Array(item._1,
            item._2,item._2))
        })
      } catch {
        case e: Exception => e.printStackTrace()
      } finally {
        sqlProxy.shutdown(clinet)
      }
    })

  })

  stream.foreachRDD(rdd => {
    val sqlProxy = new SqlProxy()
    val client = DataSourceUtil.getConnection
    try {
      val offsetRanges: Array[OffsetRange] = rdd.asInstanceOf
        [HasOffsetRanges].offsetRanges
      for (or <- offsetRanges) {
        sqlProxy.executeUpdate(client, "replace into `offset_manager`
          (groupid,topic,`partition`,untilOffset) values(?,?,?,?)",
```

```
          Array(groupid, or.topic, or.partition.toString, or.untilOffset))
        }
      } catch {
        case e: Exception => e.printStackTrace()
      } finally {
        sqlProxy.shutdown(client)
      }
    })
    ssc.start()
    ssc.awaitTermination()
}
/**
  *
  * @param key
  * @param iters
  * @param sqlProxy
  * @param client
  */
def calcVideoTime(key: String, iters: Iterable[LearnModel], sqlProxy:
  SqlProxy, client: Connection) = {
  val keys = key.split("_")
  val userId = keys(0).toInt
  val cwareId = keys(1).toInt
      val videoId = keys(2).toInt

      var interval_history = ""
      sqlProxy.executeQuery(client,"select play_interval from
        video_interval
          where userid=? and cwareid=? and videoid=?",
          Array(userId,cwareId, videoId),new QueryCallback {
              override def process(rs:ResultSet):Unit = {
                  while (rs.next()) {
                      interval_history = rs.getString(1)
                  }
                  rs.close()
              }
          })
      var effective_duration_sum = 0l
      var complete_duration_sum = 0l
      var cumulative_duration_sum = 0l
      val learnList = iters.toList.sortBy(item => item.ps)
      learnList.foreach(item => {
          if ("".equals(interval_history)) {

              val play_interval = item.ps + "-" + item.pe
              val effective_duration = Math.ceil((item.te - item.ts)
                /1000)
              val complete_duration = item.pe - item.ps
              effective_duration_sum += effective_duration.toLong
              cumulative_duration_sum += effective_duration.toLong
              complete_duration_sum += complete_duration
              interval_history = play_interval
```

```
        } else {

            val interval_arry = interval_history.split(",").sortBy(a =>
            (a.split("-")(0).toInt,a.split("-")(1).toInt))
            val tuple = getEffectiveInterval(
            interval_arry,item.ps,item.pe)
            val complete_duration = tuple._1
            val effective_duration = Math.ceil((
            item.te - item.ts)/1000)/
            (item.pe - item.ps) * complete_duration
            val cumulative_duration = Math.ceil((item.te - item.
            ts)/1000)
            interval_history = tuple._2
            effective_duration_sum += effective_duration.toLong
            complete_duration_sum += complete_duration
            cumulative_duration_sum += cumulative_duration.toLong
        }
        sqlProxy.executeUpdate(client,"insert into video_interval(
            userid,cwareid,videoid,play_interval) values(?,?,?,?) " +
            "on duplicate key update play_interval=?",Array
            (userId,cwareId,
            videoId,interval_history,interval_history))
        sqlProxy.executeUpdate(client,"insert into video_learn_detail(
            userid,cwareid,videoid,totaltime,effecttime,
            completetime) " +
            "values(?,?,?,?,?,?) on duplicate key update totaltime=
            totaltime+?,effecttime=effecttime+?,completetime=
            completetime+?",
            Array(userId,cwareId,videoId,cumulative_duration_sum,
            effective_duration_sum, complete_duration_sum,
            cumulative_duration_sum,
            effective_duration_sum,complete_duration_sum))
    })
}

/**
    * 计算有效区间
    *
    * @param array
    * @param start
    * @param end
    * @return
    */
def getEffectiveInterval(array: Array[String], start: Int, end: Int) = {
    var effective_duration = end - start
    var bl = false
    import scala.util.control.Breaks._
    breakable {
        for (i <- 0 until array.length) {
            var historyStart = 0
            var historyEnd = 0
            val item = array(i)
```

```
        try {
            historyStart = item.split("-")(0).toInt
            historyEnd = item.split("-")(1).toInt
        } catch {
            case e: Exception => throw new Exception("error
            array:" +
            array.mkString(","))
        }
        if (start >= historyStart && historyEnd >= end) {
            effective_duration = 0
            bl = true
            break()
        } else if (start <= historyStart && end >
            historyStart && end <
        historyEnd) {
            effective_duration -= end - historyStart
            array(i) = start + "-" + historyEnd
            bl = true
        } else if (start > historyStart && start <
            historyEnd && end >=
        historyEnd) {
            effective_duration -= historyEnd - start
            array(i) = historyStart + "-" + end
            bl = true
        } else if (start < historyStart && end > historyEnd) {

            effective_duration -= historyEnd - historyStart
            array(i) = start + "-" + end
            bl = true
        }
    }
}
val result = bl match {
    case false => {

        val distinctArray2 = ArrayBuffer[String]()
        distinctArray2.appendAll(array)
        distinctArray2.append(start + "-" + end)
        val distinctArray = distinctArray2.distinct.sortBy(a =>
        (a.split("-")(0).toInt,a.split("-")(1).toInt))
        val tmpArray = ArrayBuffer[String]()
        tmpArray.append(distinctArray(0))
        for (i <- 1 until distinctArray.length) {
            val item = distinctArray(i).split("-")
            val tmpItem = tmpArray(tmpArray.length - 1).split("-")
            val itemStart = item(0)
            val itemEnd = item(1)
            val tmpItemStart = tmpItem(0)
            val tmpItemEnd = tmpItem(1)
            if (tmpItemStart.toInt < itemStart.toInt &&
            tmpItemEnd.toInt < itemStart.toInt) {
```

```
                tmpArray.append(itemStart + "-" + itemEnd)
            } else {

                val resultStart = tmpItemStart
                val resultEnd = if (tmpItemEnd.toInt >
                itemEnd.toInt)
                tmpItemEnd else itemEnd
                tmpArray(tmpArray.length - 1) =
                resultStart + "-" + resultEnd
            }
        }
        val play_interval = tmpArray.sortBy(a =>
        (a.split("-")(0).toInt,
        a.split("-")(1).toInt)).mkString(",")
        play_interval
    }
    case true => {

        val distinctArray = array.distinct.sortBy(a =>
            (a.split("-")(0).toInt,
            a.split("-")(1).toInt))
        val tmpArray = ArrayBuffer[String]()
        tmpArray.append(distinctArray(0))
        for (i <- 1 until distinctArray.length) {
            val item = distinctArray(i).split("-")
            val tmpItem = tmpArray(tmpArray.length -
                1).split("-")
            val itemStart = item(0)
            val itemEnd = item(1)
            val tmpItemStart = tmpItem(0)
            val tmpItemEnd = tmpItem(1)
            if (tmpItemStart.toInt < itemStart.toInt &&
                tmpItemEnd.toInt < itemStart.toInt) {

                tmpArray.append(itemStart + "-" + itemEnd)
            } else {

                val resultStart = tmpItemStart
                val resultEnd = if (tmpItemEnd.toInt >
                    itemEnd.toInt)
                tmpItemEnd else itemEnd
                tmpArray(tmpArray.length - 1) = resultStart +
                    "-" + resultEnd
            }
        }
        val play_interval = tmpArray.sortBy(a =>
            (a.split("-")(0).toInt,
            a.split("-")(1).toInt)).mkString(",")
        play_interval
    }
```

```
        }
        (effective_duration, result)
    }
}
```

8.5 本章小结

本章首先介绍了Apache Flink是由Apache软件基金会开发的开源流处理框架，其核心是使用Java语言和Scala语言编写的分布式流数据、流引擎。Flink以数据并行和流水线的方式执行任意流数据程序，Flink的流水线运行时，系统可以执行批处理和流处理程序。此外，Flink的运行时本身也支持迭代算法。其次介绍了Flink的配置部署、管理过程，以及开发API常用功能的使用方法，并通过案例形式重点进行介绍。

8.6 课后习题

1.选择题

（1）搭建一个数据仓库可能需要使用（　　）技术。

A.Oracle　　B.MySQL　　C.Hadoop　　D.以上都是

（2）Flink用于支持离线计算的API是（　　）。

A.DataStream　　B.DataSet　　C.RDD　　D.SQL

（3）Flink流水线的本质是一种（　　）。

A.高可用机制　　B.内存管理机制　　C.持久化机制　　D.延时机制

（4）Flink on Yarn的模式有（　　）。

A.内存会话管理　　B.内存集中管理　　C.内存Job管理　　D.独立运行模式

（5）Flink可以部署在（　　）平台上。

A.local　　B.yarn　　C.spark　　D.docker

（6）Flink状态的后端存储有（　　）几种。

A.FileSystem　　B.Memory　　C.RockDB　　D.HBase

（7）Flink的并行度可以在（　　）层次设置。

A.算子级别　　B.客户端级别　　C.执行环境级别　　D.系统级别

（8）Flink的HA实现需要依赖以下（　　）组件。

A.HDFS　　B.MySQL　　C.HBase　　D.ZooKeeper

2.简答题

（1）简述Flink常用的API，并说明相关API作用的特点。

（2）简述Flink Job管理的核心技术。

（3）简述Flink的作用是什么，并简单描述Flink的功能。

参考文献

[1] 黑马程序员.Hadoop大数据技术原理与应用[M].北京：清华大学出版社，2019.

[2] 罗文浪，邱波，郭炳宇，姜善永.Hadoop大数据平台集群部署与开发[M].北京：人民邮电出版社，2018.

[3] 杨力.Hadoop大数据开发实战[M].北京：人民邮电出版社，2019.

[4] Flume官方社区网站[EB/OL].用户入门手册、用户开发手册.

[5] Kafka官方社区网站[EB/OL].社区技术手册帮助文档.

[6] PySpark官方社区网站[EB/OL].社区技术手册帮助文档.

[7] FLink官方社区网站[EB/OL].社区技术手册帮助文档.

[8] gitHub官方网站[EB/OL].Apache社区项目源码资源.